# Hélène Metzger, Historian and Historiographer of the Sciences

Is there something important to learn from the history of science about knowledge and the mind? Do habits and emotions play a significant role in science? To what extent do present concerns and knowledge distort our understanding of past texts and practices? These are crucial questions in current debates, but they are not new. This monograph evaluates the answers to these and other questions that Hélène Metzger (1889–1944) provided. Metzger, who was the leading historian of chemistry of her generation, left us unparalleled reflections on the theory, practice and aims of history writing. Despite her influence on subsequent generations of thinkers, including Thomas Kuhn, this is the first full-length monograph on her. Beginning with an overview of her life, and the challenges faced by a Jewish woman working within academia, the book goes on to discuss the most important themes of her historiography, and her engagement with other disciplines, notably general history, philosophy, ethnology and religious studies. The book also explores both Metzger's immediate legacy and the relevance of her ideas for a host of current debates in science studies. The Appendices include four of her historiographical papers, translated into English for the first time.

**Dr. Cristina Chimisso** is Senior Lecturer in Philosophy and European Studies at the Open University. She is the author of the monographs *Writing the History of the Mind: Philosophy and Science in France 1900–1960s* (2008), written with the support of an AHRC grant, and *Gaston Bachelard: Critic of Science and the Imagination* (2001).

## Science, Technology and Culture, 1700–1945
*Series Editors*

**Robert M. Brain**
*The University of British Columbia, Canada*

and

**Ernst Hamm**
*York University, Canada*

*Science, Technology and Culture, 1700–1945* focuses on the social, cultural, industrial and economic contexts of science and technology from the 'scientific revolution' up to the Second World War. Publishing lively, original, innovative research across a broad spectrum of subjects and genres by an international list of authors, the series has a global compass that concerns the development of modern science in all regions of the world. Subjects may range from close studies of particular sciences and problems to cultural and social histories of science, technology and biomedicine; accounts of scientific travel and exploration; transnational histories of scientific and technological change; monographs examining instruments, their makers and users; the material and visual cultures of science; contextual studies of institutions and of individual scientists, engineers and popularisers of science; and well-edited volumes of essays on themes in the field.

### Also in the series

**Barcelona: An Urban History of Science and Modernity, 1888–1929**
*Oliver Hochadel and Agustí Nieto-Galan*

**Pursuing the Unity of Science**
Ideology and Scientific Practice from the Great War to the Cold War
*Harmke Kamminga and Geert Somsen*

**The Enlightenment of Thomas Beddoes**
Science, Medicine, and Reform
*Trevor Levere, Larry Stewart, and Hugh Torrens, with Joseph Wachelder*

**Brainwaves: A Cultural History of Electroencephalography**
*Cornelius Borck, translated by Ann M. Hentschel*

**Hélène Metzger, Historian and Historiographer of the Sciences**
*Cristina Chimisso*

https://www.routledge.com/Science-Technology-and-Culture-1700-1945/book-series/STAC

# Hélène Metzger, Historian and Historiographer of the Sciences

Cristina Chimisso,
The Open University, UK

LONDON AND NEW YORK

First published 2019
by Routledge
2 Park Square, Milton Park, Abingdon, Oxon OX14 4RN

and by Routledge
605 Third Avenue, New York, NY 10017

First issued in paperback 2021

*Routledge is an imprint of the Taylor & Francis Group, an informa business*

Publisher's Note
The publisher has gone to great lengths to ensure the quality of this reprint but points out that some imperfections in the original copies may be apparent.

*British Library Cataloguing-in-Publication Data*
A catalogue record for this book is available from the British Library

*Library of Congress Cataloging-in-Publication Data*
Names: Chimisso, Cristina, author.
Title: Hélène Metzger: historian and historiographer of the sciences/ Cristina Chimisso.
Description: Abingdon, Oxon; New York, NY: Routledge, 2019. |
Series: Science, technology and culture, 1700–1945 |
Includes bibliographical references and index.
Identifiers: LCCN 2019036205 (print) | LCCN 2019036206 (ebook) |
ISBN 9781138210394 (hardback) | ISBN 9781315455372 (ebook)
Subjects: LCSH: Metzger, Hélène. | Historians of
science–France–Biography. | Women philosophers–France–Biography. |
Science–History. | Science–Philosophy. | Science–Historiography.
Classification: LCC Q143.M48 C45 2019 (print) | LCC Q143.M48 (ebook) |
DDC 507.2/02 [B]–dc23
LC record available at https://lccn.loc.gov/2019036205
LC ebook record available at https://lccn.loc.gov/2019036206

Typeset in Sabon
by Deanta Global Publishing Services, Chennai, India

ISBN 13: 978-1-03-208303-2 (pbk)
ISBN 13: 978-1-138-21039-4 (hbk)

# Contents

# Acknowledgements

My interest in Hélène Metzger started long ago. It was to work on her philosophy that the American Academy of Arts and Science awarded me the May and George Sarton fellowship back in 1999. Thanks to the American Academy, I was able to spend time in the Houghton Library at Harvard University, reading Sarton's papers, which include his correspondence with Metzger and with a number of philosophers and historians of her milieu. My research at that time resulted in a long article on her, published in *Studies in History and Philosophy of Science* in 2001. Since then, while I worked on other projects, she has been in the background, making an appearance in my book *Writing the History of the Mind* (Ashgate 2001), for which I received the support of the Arts and Humanities Research Council. Almost every year, I present Metzger's historiography to the graduate students of the Department of History and Philosophy of Science at the University of Cambridge, as part of the cycle of seminars 'Aims and Methods of Histories of the Sciences', organised by Nick Jardine. While I was having coffee with him before one of these seminars, Nick convinced me that Metzger deserved a full monograph. He also very kindly discussed the planning of this project with me, and indeed commented on the full draft. I cannot thank him enough. I would also like to express my gratitude to Gad Freudenthal, my comrade in all things Metzger, as he himself put it. He generously sent me copies of the documents regarding Metzger that he had gathered. Over the years, he had also contacted Metzger's friends, colleagues and relatives, who answered his questions; he sent me copies of their replies, which shed light on Metzger's life and inclinations. I understand that he has now donated those documents and letters to IMEC (Institut Memoirs de l'édition contemporaine). I even bothered him when I had doubts on how best to translate some of Metzger's expressions, and he always answered very promptly and helpfully. Carolyn Price advised me on the recent philosophical literature on emotions. I presented parts of this book as work in progress at various conferences and seminars, including at the HOPOS conference in Groningen (2018), at the annual conference of the British Society for the History of Philosophy in Sheffield (2017) and at the symposium on Mach, Duhem and French philosophy of science, part of the Ernst Mach conference at the University of Vienna (2016).

# Introduction and biographical notes on Hélène Metzger

Is there something important to learn from the history of science about the production of knowledge and the human mind? Is science a purely rational enterprise, or do habits and emotions play a significant role in it? To what extent do present concerns and knowledge distort our understanding of past texts and practices? These are live questions in current debates, but they are not new. Indeed, many current scholars engage with the ideas of past philosophers and historians in order to answer them. Scholars who are involved in the cluster of projects labelled 'historical epistemology', including among others Ian Hacking, Arnold I. Davidson and Hans-Jörg Rheinberger, have acknowledged their debt to the tradition, also called historical epistemology, which flourished in France in the interwar period and beyond. However, the latter tradition is often reduced to the works of Gaston Bachelard and Georges Canguilhem. In reality, 'classic' historical epistemology, as distinct from current projects, developed thanks to a number of diverse scholars who debated with one another the historical nature of knowledge, indeed of our mental categories, the role of reason, the imagination and religion in the history of science, and the demarcation between genuine knowledge and other systems of beliefs. Several of these scholars will appear in the present book, including Léon Brunschvicg, Lucien Lévy-Bruhl, Émile Meyerson, Alexandre Koyré, and naturally Gaston Bachelard and Georges Canguilhem. However, those and other scholars, such as Henri Berr, Abel Rey, Pierre Duhem and George Sarton, will be considered in relation to the historian and philosopher Hélène Metzger, née Bruhl (1889–1944). She is the author of classic monographs in the history of chemistry and is highly respected for her innovative and perceptive ideas on the historiography of the sciences. However, undeservedly, her fame has been limited to a relatively small group of scholars, despite the fact that she is one of the scholars whom Thomas Kuhn cited as his inspiration in the Introduction of *The structure of scientific revolutions*. Elsewhere, he presented her work as an important example of the 'new historiography', as we shall see in Chapter 8 (Kuhn, 1996 [1962]); (Kuhn, 1977).

In Part I shall discuss Metzger's historiography and epistemology. We shall see the way in which she tackled the problem of anachronism, both in her theoretical essays and in her histories of chemistry and crystallography. She delved into past scholars' uses of concepts such as analogy, substance and evolution; and she emphasised the vast material, conceptual and meta-physical differences between the world of past scholars and that of modern readers. Like her fellow historical epistemologists, she aimed to study the mind, and in particular the minds of past scholars and, as she put it, of the educated public around them. In order to do this, the traditional history of science focussed on theories in their final form, or even updated for the modern reader, was of little use to her. Rather, she aimed to study 'science in the making', in the double meaning of the moment in which intuitions and ideas emerge, and the historical early phase of the development of a discipline. At a time when philosophers like Karl Popper thought that the 'context of discovery' was of no interest to philosophers, she argued that it is precisely the discovery phase that will best reveal to us how knowledge is acquired. She thought that, by maintaining the focus on science in the mak-ing, we can discover the role that habits and emotions play in the formation of knowledge. Her view was at odds with that of philosophers and histo-rians in her milieu, who regarded science as eminently rational. They were equally mystified by her declared use of what she called 'active sympathy' in her work as a historian.

Part II will discuss Metzger's engagement with disciplines and scholars around her. I shall place her in the crucial debate of her time and beyond that saw the proposers of a discontinuous model of history, which included episte-mological breaks, face the proposers of a continuous model. We shall see that she maintained a nuanced and balanced position, which corresponded neither to Brunchvicg and Bachelard's discontinuist, nor to Duhem's continuist view of history of science. Philosophy was crucial for Metzger, both as a historical method, which she indeed called 'philosophical', and as the aim of history of science, which for her is the study of the human mind. I shall discuss how she engaged with the philosophies of her time. In interwar France, history of science and ethnology were closely linked to philosophy. It is in this con-text that we should understand her engagement with Lévy-Bruhl's concept of primitive mentality. Lévy-Bruhl was the founder of the Institut d'ethnologie at the University of Paris, but was also a professor of history of philosophy at the same institution. Philosophers, in addition to sociologists, psychologists and historians, debated Lévy-Bruhl's proposal, and so did Metzger, who inci-dentally was Lévy-Bruhl's niece and close to him. Lastly, I shall analyse her view of the relationship between science on the one hand and theology and religion on the other, as it was quite original in her milieu. She emphasised the role that religion has played in science, and indeed the role that science has played in religion. Moreover, in the last years of her life, she was writing a book in which she discussed the relationship between science, human will and religion. She wrote an extended plan of this work in Lyon, as she was fleeing Nazi-occupied Paris. Her critique of totalitarianism appears to be an important motivation behind her planned monograph.

In Part III, I shall focus on Metzger's impact on her world and on our own. Although she is often presented as a mentee, she lectured at the Sorbonne's Institut d'histoire des sciences and at the École pratique des hautes études, and had places of responsibility at the Centre de synthèse and at the International Committee for the History of Science, renamed Academy for the History of Science in 1932. Above all, she was a highly respected scholar, and the impact of her work is well documented, notably in Bachelard's and Canguilhem's discussions of her ideas and by their reliance on her scholarship. By contrast, her impact on our current debates has regrettably been very limited, although several scholars have argued that her ideas should play a greater role in present-day philosophy and historiography. Lamentably, neither Kuhn's praise and use of her work, nor the current reception of historical epistemology have facilitated the knowledge of her work in our world. I shall propose that some of her ideas, with all inevitable modifications, can still enrich and stimulate our historiographical and philosophical reflections. Of particular interest are her critique of anachronism, her focus on 'science in the making', her study of analogical thought in history of science, her discussion of emotions and habits in science, and her defence of the use of sympathy in historical knowledge.

One important reason why knowledge of her work has not been as enduring and widespread as one could expect is her gender. In fact, her gender created many obstacles to her academic life, and had a negative impact on the reception of her work even in her own lifetime. As mentioned, she lectured and had posts of responsibility, but she never had a full-time and paid academic job. Although she has been included in the number of twentieth-century women who were 'amateurs by choice' (Pomata, 2013), in fact she was no amateur, and her lack of an academic post was never a choice. She intensely desired to be a full member of academia. A few months before her thirty-seventh birthday,[1] she wrote to the historian of science George Sarton: 'I shall conquer a post somewhere. But I am for ever stopped in my efforts' (Freudenthal, 1990a, Metzger to Sarton, Letter of 22.4.1926, pp. 254–255). She was aware of her own value, and of the nature of the obstacles on her path. At the time, she had already published three books: two monographs on the history of science, *La genèse de la science des cristaux* (Metzger, 1918), and *Les doctrines chimiques en France du début du XVIIe à la fin du XVIIIe siècle* (Metzger, 1969 [1923]), and a philosophical volume, *Les concepts scientifiques* (Metzger, 1926c). A full academic post in French academia was not a very realistic prospect for a woman. In 1930, in the whole of France there were only six women holding university positions (Clark, 1937, p. 48). At the University of Paris there were two, both in the faculty of science, one of whom was Marie Skłodowska-Curie. She had been appointed to her late husband's chair in 1908, when she had already won her first Nobel Prize (with her husband Pierre and Henri Becquerel); she was awarded her second in 1911 (Anonymous, 1934); (Dry, 2003, p. 73). Metzger did not comment on the general exclusion of women from academia, but rather on her own lack of adequate qualifications. However, she was aware that the reason why she had not received the appropriate education for an academic career[2] was that she was a woman. She claimed that her father just followed the ideas of his time, and did not plan for his daughters to have independent professions. Incidentally, her younger half-brother, Adrien Bruhl, received a more prestigious education, and eventually became the Dean of the faculty of letters of the University of Dijon. The type of education that Metzger's father chose for her and her sister created an obstacle, not so much to an independent profession, as to an academic career. She nevertheless went into higher education, obtaining a *diplôme d'étude supérieures* in crystallography (Freudenthal, 1990b).[3] Metzger understood very well what her father's plans for her daughters were. Rather than allowing them to have the type of education that would open all doors, she explained to Sarton, he provided them with dowries, so that they could marry young men 'of intellectual and moral value' but with only a modest income. The model she described has been subsequently confirmed as a general pattern among French Jews of her time: the marriage between a well-off woman and a promising intellectual without fortune. The exemplar of this marriage, in Christopher Charles' presentation, is that of Émile Durkheim to wealthy Julie Dreyfus (Charle, 1984). The Bruhls, as diamond, pearl and precious stone merchants,[4] were well-off. Their extended family had already celebrated one of these marriages. Alice Louise Bruhl, Hélène's paternal aunt, had married Lucien Lévy, a promising intellectual. With his marriage he became Lucien Lévy-Bruhl, and it is with this name that we know the professor of history of philosophy at the Sorbonne, co-founder of its Institute of Ethnology. Other Jewish intellectuals in Metzger's milieu made similar marriages. A case in point is Henri Berr (1863–1954), the director of the Centre de synthèse, an independent organisation for the study of history with which Metzger was closely involved.[5] I do not know how disappointed the family of Henri Berr's wife were at his failure to secure a chair at the Collège de France, despite Henri Bergson's support (Gemelli, 1987, p. 231). It is however certain that without his wife's wealth and connections, Berr could not have created the prestigious Centre de synthèse, which attracted highly regarded scholars and the support of prominent politicians (Chimisso, 2001). Another similar case is that of the Sorbonne professor of history of philosophy Léon Brunschvicg, who played an important role in Metzger's development as a scholar: he was of humble origins, but married Cécile Kahn, the daughter of a wealthy merchant.[6]

Hélène's own marriage followed that pattern, and indeed her father's plans: in 1913, she married a promising intellectual, Paul Metzger. After studying law, Paul started his career at the University of Lyon as a legal historian, working, like Hélène, on the eighteenth century. His monograph was published the year of their marriage (Metzger, 1913). Hélène, described as a 'feminist' by Suzanne Delorme (Delorme, 1983) and 'socialist, or almost' by herself (Freudenthal, 1990a, Metzger to Sarton, Letter of 22.4.1926,

p. 254), did not abandon her studies, but rather turned to the history of the science she knew best, crystallography. In 1914, 'M^me H. Bruhl-Metzger' (she would later drop her maiden name) published an article on 'Buffon's curious theory of double refraction' in the *Bulletin de la Société française de Minéralogie*, a scientific journal (Bruhl-Metzger, 1914). During the First World War, she wrote the manuscript of *La genèse*. Already in her first book-length work, she faced the historiographical concerns on which she was going to reflect throughout her life. Notably, she aimed to understand the different ways in which past scholars conceived of nature and of their objects of study, and strove to avoid projecting her own scientific knowledge and assumptions on her sources. I shall discuss her view on these issues in detail in Chapter 1 'Anachronism and scientific concepts'. *La genèse* also already displays her life-long focus on the early phases of a science, and on science 'in the making', a theme that I shall discuss in particular in Chapter 2 'The study of science in the making'. She wrote that in order to understand a science like crystallography, the examination of its definitive form will not suffice. She indeed warned her readers that the impression they may have of the 'sudden appearance' of a fully formed system of theories and practices is illusory. The war years were not just a period of study for Metzger; they also brought tragedy, as they did for many others. Her husband was an early casualty of the war, and Hélène found herself a widow at twenty-six years of age, only sixteen months after her wedding. Reportedly, she wore black for the rest of her life (Delorme, 1983). There were no reasons for her to stay in Lyon; she moved to Paris, and set up her home in rue de Pauquet, in the 9th arrondissement, where she lived until she was forced to flee Paris due to the German occupation. She believed that her work was of the right standard for a doctorate, and, to this end, she gave it to read to her former professor of crystallography. Metzger's focus on the 'prehistory', as it were, of crystallography, and her interest in minor scholars and on ideas that modern scientists regarded as wrong and indeed peculiar, did not impress the scientist. He judged her research pointless, and could not imagine that the subject matter would interest anyone. But it did: as she discovered, among philosophers and historians there were serious historians of science (Freudenthal, 1990a, Metzger to Sarton, Letter of 20.6.1922, p. 250). It is plausible that her uncle Lucien Lévy-Bruhl introduced her into the milieu of philosophers. She reported that Gaston Milhaud (1858–1918), Sorbonne professor of history of philosophy in its relationship with the sciences, was extremely positive and encouraging about her work, which indeed afforded her a doctorate. However, as she did not hold the type of degree (*licence*) required in order to pursue a 'full' *doctorat d'État*, she was granted a *doctorat d'université*, a shorter degree that was overwhelmingly pursued by foreigners and women, and that did not open the doors to an academic career.

Milhaud died in 1918, the year in which *La genèse* was published as a book. At this stage, however, Metzger was in a congenial environment. She attended the lectures of Léon Brunschvicg (1869–1944), a professor 'without chair' until 1927, when he obtained Lévy-Bruhl's chair on the latter's retirement (Guigue, 1935, pp. 32, 244); (Charle, 1986, pp. 44–45). He eventually became one of the most powerful professors in Paris (Bourdieu, 1988 [1984], p. 93). In her obituary of Brunschvicg, Suzanne Delorme emphasised that he was 'particularly interested' in the work of 'the historian of chemistry Hélène Metzger' (Delorme, 1947–48, p. 519).[7] As I shall discuss in Chapter 5, Brunschvicg's historicised Kantianism, and his study of the mind through the history of science, were important in Metzger's philosophical development and lent support to her historiographical views. On the other hand, she neither pursued the grand narratives that were Brunschvicg's speciality (Brunschvicg, 1912); (Brunschvicg, 1922), nor did she share his fully rationalistic view of science and its progress. In fact, she assigned a role to emotions and habits in the development of science, which I shall analyse in Chapter 3. Along with Brunschvicg, she considered André Lalande her philosophy teacher (Delorme, 1983). Lalande (1867–1963), professor of philosophy at the Sorbonne, and editor of the standard Dictionary of Philosophy (Lalande, 1999 [1926]), took over from Milhaud as Metzger's mentor. He must have thought that Metzger should receive prizes for her work, because he supported her for two of them. He recommended *Les doctrines chimiques* for the 1924 Binoux Prize of the Académie des sciences morales et politiques for the best philosophy or history book published in 1923, with two letters to the Academy's secretary and to another member.[8] The latter letter emphasises the originality and value of Metzger's book, but at the same time, somewhat patronisingly, depicts her as a war widow who has devoted herself to research as a way of coping with her bereavement. Lalande also mentioned that Metzger was Lévy-Bruhl's niece, and that he hoped that this prize would encourage her. Metzger was awarded the prize, which gave a boost to her reputation; as she remarked, within a week everybody seemed to take her work seriously, rather than discount it as a useless fantasy (Freudenthal, 1990a, Metzger to Sarton, Letter of 6.12.1924, p. 252); (Freudenthal, 1990b, p. 200). She also entered for the Bordin prize for philosophy of the Académie des sciences morales et politiques, with a manuscript that she had written following the advice of 'her professors' (Metzger, 1924, Letter of 23.10.1924). The submission was anonymous, but once again, it was Lalande who wrote the report on it, and Metzger won the prize. His report to the Academy presented the then manuscript as an original and erudite work, without glossing over its shortcomings, including what he judged its imprecise use of philosophical terminology, perhaps unsurprisingly from the editor of the philosophical dictionary of reference. It is without doubt that Lalande gave Metzger his unwavering support, although his tone in his reference letters and indeed report is such that, despite his clear praise, the reader cannot imagine that he is presenting somebody whom he considers a peer. In any case, her prize-winning manuscript was published in 1926 as *Les concepts scientifiques*, Metzger's most philosophical book, and her most 'striking', according to

Charles Singer (Singer, 1946).[9] These two prizes sanctioned her value as a scholar in the eyes of many, but by no means all.[10] In the year in which *Les concepts scientifiques* came out, she wrote to Sarton of the 'hurtful' rumour according to which she was an extremely rich woman ('a Carnegie's daughter') – which, she implied, she was not – who bought herself an 'aristocrat', that is her academic husband. She was particularly saddened that such rumour came from her late husband's friends, who treated her as an interloper in their academic world. They, she wrote, 'believe[d] academics superior to anybody' and *a priori* despised any 'historical, philosophical, or scientific work' that was not written by an academic. They suggested, to use Pierre Bourdieu's terminology, that she aspired to join the 'State nobility' by marriage rather than merit (Bourdieu, 1998). The problem, as she saw it in 1926, was that since she had 'no official position', she was classed 'in the category of amateurs' (Freudenthal, 1990b, Metzger to Sarton, Letter of 22.4.26, pp. 254–255).

She also had strong, though difficult, links with the philosopher Émile Meyerson. As we shall see in Chapter 7, Metzger has been repeatedly classed as his disciple. Indeed, he appears to be the most cited 'influence' on her philosophy. The cosmopolitan and polyglot Meyerson was an independent scholar, but his social network was impressive, and included the most highly regarded philosophers and scientists.[11] He was a close friend of both Léon Brunschvicg and Lucien Lévy-Bruhl, notwithstanding their philosophical differences.[12] It was therefore almost inevitable that Metzger should meet him, considering that they also shared similar philosophical concerns and a background in chemistry. Meyerson, thirty years her senior, willingly took the role of teacher, but, at least in Metzger's perception, he regarded her as an eternal pupil, never as a fellow scholar, no matter how much she affirmed her intellectual independence. In a letter to Sarton, written while she was working on her monograph *Newton, Stahl, Boerhaave* (Metzger, 1974 [1930]), she placed a telling exclamation point at the end of a sentence reporting that Meyerson told her that the best chapter of her manuscript was the one that he had closely supervised (Freudenthal, 1990b, Metzger to Sarton, Letter of 10.8.1925, p. 253).[13] Nevertheless, when the book came out, there was a dedication to him, though tucked at the end of the Introduction, unlike the dedication to Lévy-Bruhl of *Les doctrines chimiques*, which occupies the page after the title. When she met Meyerson shortly after the publication of *Les concepts scientifiques*, she was hurt as he failed to mention her book, while asking her to compile the index of his own book 'on a similar topic'. She showed impatience with his attitude: she wrote that if he so wished, she could declare herself his pupil, although all she had published was written outside his influence. She added that she refused to be the slave of even the greatest philosopher, as she had a brain (Freudenthal, 1990a, Metzger to Sarton, Letter of 14.4.1927, p. 255). She even decided to affirm her independence from him in a public talk. At the first conference that the International Committee for the History of Science organised (Paris, 1929), she gave a talk on his philosophy, later published in *Archeion*. In this talk, she emphasised that she 'unfortunately' could not say that he was her teacher, because she 'studied his books after doing a great deal of work and publishing two volumes'; and because the scope of her research was different from that of his epistemological work (Metzger, 1987 [1929], p. 95). Her remarks in the letters to Sarton and in the Paris talk, however, almost pale into insignificance when compared with the letter that she wrote to Meyerson in 1933. While expressing her admiration for him, which, she added, was particularly valuable because it did not come from 'subordination', she reminded him that she was not a child, but a mature woman of forty-three years of age. She wrote:

> ... do not try to modify me, change me, educate me, deform me, or, in a purely school-like manner, be my 'schoolmaster'. I have always considered as a 'potential' enemy whatever individual (a parent, a teacher, a physician, etc.) who is invested with a patch of authority and who wants to use their prestige in order to impose upon me their ideas or way of seeing things ... I am an inborn democrat.
>
> (Chimisso and Freudenthal, 2003, Metzger to
> Meyerson, p. 490)[14]

In Paris, at first, she had been in contact mainly with philosophers; indeed with the most important philosophers of science of the time. However, she soon forged an important link with the nascent discipline of history of science when George Sarton (1884–1956) sought her collaboration with *Isis*, the journal that the Belgian historian of science had founded and then taken with him to the United States. In 1921, when she and Sarton started a regular correspondence, she was not even aware that *Isis* existed (Freudenthal, 1990a, Metzger to Sarton, Letter of 20.7.1921, pp. 147–148).[15] Sarton had been in contact with French scholars with an interest in the history of science, including Henri Bergson, Pierre Duhem, Léon Brunschvicg, Henri Poincaré, Henri Berr and Abel Rey, with mixed results. He had a positive response from Metzger, who between 1922 and 1933 regularly wrote reviews for *Isis* of the most prominent French books in history and philosophy of science and related fields, including ones by Léon Brunschvicg, Lucien Lévy-Bruhl, Émile Meyerson, André Cresson and Gaston Bachelard.[16] *Isis* also published her original articles, which she then incorporated into *Les doctrines chimiques* and *Newton, Stahl, Boerhaave* (Metzger, 1922b); (Metzger, 1926a); (Metzger, 1927c); (Metzger, 1974 [1930]). *Newton, Stahl, Boerhaave* came out in 1930, at the same time as her popular history of chemistry *La chimie*, translated a few decades later as *Chemistry* (Metzger, 1930); (Metzger, 1991 [1930]).

Metzger collaborated with *Isis* and regarded Sarton not only as a fellow historian of science, but also as a friend. However, their approaches to the history of science differed considerably. Whereas Sarton aimed to offer a grand narrative of history of science, especially in his major work (Sarton, 1927–1947), Metzger kept her focus on the seventeenth and eighteenth centuries and on specific disciplines and practices. More importantly, as we shall see in particular in Chapter 5, she regarded philosophy as playing a crucial role in the objectives as well as the method of history of science. Sarton did not share her philosophical views; on the other hand, he was far more interested than her in the creation of bibliographies and accumulation of facts. A historian who shared Sarton's focus on bibliographies and inventories was Henri Berr.[17] As we shall see in Chapter 4, Berr intended the Centre de synthèse as the instrument for the promotion of history as a science, rather than as 'a literary genre' (Berr, 1911, p. 232). His emphasis was on empirical research, fact and document gathering, and rigorous definition of terms in the form of a dictionary. The Centre de synthèse also became a hub for historians of science when the historian of science Aldo Mieli fled Italy in 1928 due to his well-founded fears of the Fascist police (Chimisso, 2011). The Centre de synthèse provided a home for his journal *Archeion* (originally *Archivio di storia della scienza*, and now published as *Archives internationales d'histoire des sciences*) and for the International Committee (later Academy) for the history of science, which he created in 1928. At the Centre, Mieli also founded and directed the Unit for the history of science (Section d'histoire des sciences). From the Unit's foundation in 1930, Metzger, as one 'of the most active members' both of the Committee and of the new-born Unit, was appointed its secretary (Section d'histoire des sciences, 1930, pp. 55–56). The International Committee co-ordinated national groups that promoted the history of science. Its first executive committee included Abel Rey, George Sarton, Henry E. Sigerist, Charles Singer, Karl Sudhoff, Lynn Thorndyke, and Mieli as director (Mieli, 1932); (Corsini, 1950). Elections to establish membership were called. Metzger, to her great surprise, was soon elected as a life-long member (Comité international et Centre international d'histoire des sciences, 1928); (Mieli, 1928, Letter to Sarton of 3.12.1928).[18] Despite not feeling up to the honour, she also hoped that the election would lend her some authority, especially as Berr had just asked her to work on filing documents (Freudenthal, 1990a, Metzger to Sarton, Letter of 13.2.1928, p. 258). She was later to become the Committee's librarian, administrator and treasurer (Mieli, 1930, p. 125); (Académie internationale d'histoire des sciences, 1935).

The Committee put her in touch with the international community of historians of science. She took part in the conferences organised by the Committee as well: Paris in 1929; London in 1931; Portugal in 1934; Prague in 1937. Although Mieli hardly ever mentioned her in official documents pertaining to these conferences, it seems that she helped organise them. In fact, the local organiser of the Portugal conference recorded his thanks to her for her help (Anonymous, 1936). In 1935 she represented the Committee, now called Academy, at the *Settimana della scuola di storia delle scienze*, chaired by the historian of science Federigo Enriques in Rome, a city in which Mieli would not have felt safe (Académie internationale d'histoire des sciences, 1935). At the Prague conference, when Abel Rey failed to turn up, Metzger had to give an address on behalf of the French historians of science and extend the official greetings of the French Republic to the Czechoslovak Republic (Freudenthal, 1990a, Metzger to Sarton, Letter of 1.1.1937). Without doubt Metzger did a lot of work behind the scenes on behalf of the Committee. Another piece of evidence of her work is available regarding the problems with the 1934 conference, which was originally planned to take place in Berlin. When the Nazis seized power, the Committee agreed to move it. A disagreement, however, emerged in the manner of doing so. Metzger had to use all her diplomacy to avert a split in the Committee, as Mieli was happy to announce that the conference in Berlin was postponed; the German organiser, Paul Diepgen, was exerting pressure to announce that a conference in Germany would probably take place in the next few years; and Charles Singer and Dorothea Waley Singer insisted from London that the Committee should publicly deplore the political events in Germany. Metzger wrote to Dorothea Singer in English, a language that Mieli did not know, and indeed expressed some sharp opinions about him. In the end, she managed to make everybody agree on an acceptable wording.[19]

At the beginning of the 1930s, Metzger appeared to have gained official positions in the respected Centre de synthèse, and she had also gained an international profile. However, her male colleagues struggled to see her as a peer. Indeed, in 1932, she reported that at the annual meeting of the International Committee she had been reproached for not dedicating herself entirely to filing and cataloguing (Metzger, 1932d, Letter to Sarton of 2.6.1932). At any rate, the Centre became her most stable academic home. *Archeion* progressively replaced *Isis* as the journal to which she contributed book reviews, and indeed articles. Most importantly, *Archeion* published the talks on historiography that she gave at the Centre. It was there that she presented her ideas about the aims and methods of history. It was not the best audience; many members, including Henri Berr and Aldo Mieli, had rather different views of history and history of science in particular, as we shall see in Chapters 4 and 5. They could not share her opposition to history as a record of events, or in the case of history science, as a record of fully formed theories and of discoveries as presented when extensively tested. She was always interested in the processes that lead to the emergence of new ideas, and over the years she increasingly presented as the main aim of history of science the understanding of thought processes of past scholars. Although she did not share the Centre's focus on dictionaries and bibliographies, she worked on entries for the Centre's planned dictionary; we have her contributions regarding the terms 'alchemy' and 'atom', the former published in her lifetime, and the latter posthumously (Metzger, 1938a, Metzger, 1947a). When, in 1932, Abel Rey founded the Institut d'histoire des sciences at the University of Paris, Metzger was from the start invited to lecture there, as we shall see in Chapter 7. From 1934, the Institut d'histoire des sciences had also its own journal, *Thalès*, edited by Rey, and a book series published by Hermann. Metzger contributed to *Thalès* extracts from her lectures at the Institut and an original article on the role of precursors in the history of science.[20] She was also the author of the second volume in the book series – just after the first authored by Rey himself – which was drawn from her lectures on Lavoisier's chemical philosophy (Metzger, 1935b). In his (unsigned) introduction, Rey emphasised that Metzger's lectures perfectly shared the Institut's aims, as they brought to light the philosophical ideas implicit in science, and in so doing explained its development (Metzger, 1935b, p. 1). The Institut relied on academics working for it. Most of them, however, had academic jobs; this is probably why Rey singled out Metzger when he extended his thanks for her voluntary contribution (Rey, 1936, p. 341). Nevertheless, she may have thought that the Institut was the institution where she could obtain the 'post' that she intended to 'conquer' in 1926. She may have regarded it as her teaching position alongside the research situation at the Centre de synthèse. Alas, just as the Centre de synthèse, the Institut had disappointments in store for her. In 1937, she wrote that she had no influence there, and the post of secretary that had been promised to her had in fact gone to 'young' Pierre Ducassé (he was sixteen years her junior), who fought a 'silent war' against her (Freudenthal, 1990a, Metzger to Sarton, Letter of 1.11.1937, p. 260).

From 1934 Metzger started attending Alexandre Koyré's lectures at the École pratique des hautes études, and immediately recognised Koyré's approach to history of science as similar to her own. The École pratique was very different from the Sorbonne in scope and type of students. It was open-access and the only type of examination was the submission of a thesis, typically after two or three years of attendance. Lecturers did not need any specific qualification; they were appointed on academic merit alone (Koyré, 1986, pp. 6–17, 43ff). In 1937–38, Koyré spent most of the academic year at the University of Cairo, and arranged for Metzger and Alexandre Kojève to replace him (Koyré, 1986, 43ff). Metzger's lectures on English commentators on Newton resulted in a manuscript that earned her a diploma of the École pratique and was published with the title *Attraction universelle et religion naturelle chez quelques commentateurs anglais de Newton* (Metzger, 1938b). Religion was at the core of her new publication also because the lectures on which it was based were delivered for the department of religious sciences, which had been created in 1886, the year after the study of theology had been removed from state universities, in line with the separation between church and state. Nevertheless, Metzger found it amusing that her diploma was in religious sciences: 'here I am a theologian!' she joked in a letter to George Sarton (Freudenthal, 1990a, Metzger to Sarton, letter of 1.1.1937, p. 260). Metzger never obtained a traditional, full-time academic job, which was all but beyond her reach in the social and political reality of interwar Paris. Nevertheless, she 'conquered' a series of very respectable places in the world, at the Sorbonne, the Centre de synthèse, the

International Academy for History of Science, and the École pratique. In these places, however, she faced a string of disappointments and sometimes humiliations. She had to bear the constant requests for what she regarded as clerical work at the Centre de synthèse. She was not rewarded with the post of secretary at the Institut d'histoire des sciences, despite her dedication to the teaching there. She was very aware that she was often seen as an eternal junior scholar, or in a position of support to other people's research. However, as we shall see in Chapter 7, her own work was in fact respected, cited and discussed by the most important philosophers of science of the following generations, notably Gaston Bachelard and Georges Canguilhem.

Her life, and the world of philosophers and historians that she had joined as a result of much work and commitment, started disintegrating in the late 1930s. The first to leave was Aldo Mieli. He had been spied upon by the Italian police since his youth for his political activism in the socialist party. After Mussolini took power, the Fascist secret police kept an even closer eye on him, as a 'dangerous socialist' and as a homosexual (*Bollettino delle ricerche*, 1930). When, in 1939, Italy passed the so-called racial laws, and signed an alliance with Nazi Germany, Mieli, who was an Italian national and Jewish, had further reasons to flee once again. He did not discuss his plan with Metzger, despite their ten-year-long collaboration at the Centre de synthèse. She thought he had gone to give some conference papers in Argentina. When he failed to return, she wrote to him, but he did not reply, and she was confused by his silence (Freudenthal, 1990a, Metzger to Sarton, Letter of 16.5.1940, p. 262).[21] Many Jewish intellectuals in Metzger's circle attempted to leave. Among those who succeeded was Alexandre Koyré, who moved to New York and co-organised the École libre des hautes études.[22] Metzger, however, stayed in Paris until 1941. Gad Freudenthal has seen her choices as consistent with those of that part of French Jewry who had been long established in France, and to whom the rehabilitation of Captain Dreyfus had given a sense of security. Metzger, he remarks, considered herself completely French (Freudenthal, 1990b, p. 202). His view is confirmed by Metzger's letters during the war, in which she repeatedly mentioned her husband's death 'for France' in the First World War. It is also confirmed by her behaviour during the occupation. As Freudenthal also reports, citing a letter written to him by Suzanne Delorme, Metzger, meeting the latter in a café in occupied Paris, expressed loudly her opinions about the Germans, the collaborators and Pétain's supporters, without concerns for her safety or indeed that of her companion (Freudenthal, 1990b, p. 197); (Delorme, 1983). She then moved to Lyon, where she had family, and kept working on two projects. One of them, which she mentioned several times between 1940 to 1942, was on light and chemical theory from Newton to Fresnel (Freudenthal, 1990a, Metzger to Sarton, Letters of 10.10.1940, 7.9.1941, 12.8.1942, pp. 263–265), and appears to have been lost. Of the other, *La science, l'appel de la religion et la volonté humaine*, we have an extended plan. *La revue philosophique* posthumously published its Conclusion; subsequently it came out as a short book curated by her brother Adrien Bruhl (Metzger, 1947b, Metzger, 1954). From Lyon, she wrote that the lack of books, paper and a typewriter made her work difficult (Freudenthal, 1990a, Metzger to Sarton, Letters of 7.9.1941, 12.8.1942, pp. 263–265); she must have been further saddened by the news that all the books that she had left in her Paris flat had been stolen (Sarton, 1942, Letter to Mieli of 10.8.1942). News of arrests and executions must have also intensified the sense of danger. In 1942, Sarton wrote to her that he had tried to bring her to the United States, but without success. In that letter, he encouraged her by saying that the happy days would return. She was arrested on 8 February 1944, interned, and then deported to Germany. Officially, her death has been set on 12 March 1944, at Auschwitz (Diatta, 1983).

## Notes

1 She was born on 26 August 1889, as the date on her passport shows. The photocopy of her passport has been kindly sent to me by Gad Freudenthal.

2 The route to an academic career started early with the choice of secondary school. This had to be a *lycée*, ideally one of the prestigious Parisian ones, followed by the *baccalauréat*, which opened the doors to a full university degree (*licence*). A *licence* in turn was needed to sit for the *agrégation*, the state examination that was, and still is, required to hold a teaching position in a public institution. Hélène's education fell short almost from the beginning: her secondary school was not a *lycée* and only afforded her a *brevet supérieur*. The lack of a *baccalauréat* blocked her path to a *licence*.

3 Metzger deeply resented the laws that prevented people who, like her, had not studied Latin and Greek in school from obtaining full academic degrees and pursuing academic careers. However, she thought that academics were on the whole open-minded, and not responsible for those restrictions; the problem was the government in power, and the fact that France was the most 'reactionary' country (Freudenthal, 1990a, Metzger to Sarton, Letter of 20.6.1922, p. 249).

4 Bruhl et Cie, 46 rue La Fayette (Parix IX) (Freudenthal, 1990b, p. 205) See Freudenthal's article also for further information about the Bruhls' origins, and Hélène's early life, which I do not present here. Freudenthal's article, and the documents that he gathered and very kindly passed on to me, were fundamental for the present biographical sketch.

5 Giuliana Gemelli has drawn a parallel between Metzger's and Berr's social positions, including their respective marriages (Gemelli, 1987).

6 For the details of Brunschvicg's family see Charle (1986), pp. 44–45.

7 Brunschvicg was also more open-minded about women's role in society than many men of his time were. He was a member of the Ligue d'électeurs pour le suffrage des femmes (which his wife Cécile Brunschvicg, née Kahn, chaired from 1924), and of the executive committee of the Ligue des droits de l'Homme (Charle, 1986); Cécile was a women's rights activist, and served in the 1936 Blum government as under-secretary of state for education.

8 This second letter is addressed to 'Monsieur le Recteur'. Gad Freudenthal, who has kindly sent me the copies of these letters, held at the Archives of the Institut de France, thinks that the addressee should be Louis de Launay. I wonder

whether another guess could be Paul Appell, another member of the panel, who was recteur de l'Académie de Paris.

9  In this book, as we shall see in Chapter 1, Metzger also employed Lalande's concept of 'dissolution' for her reflection on the concept of evolution and its variations (Lalande, 1899).

10  My copy of *Les concepts scientifiques* carries an inscription by Hélène Metzger 'A M Gustave Le Bon, hommage respectueuse de l'auteur, H Metzger'. When I bought it, the pages were uncut. Le Bon was eighty-nine when *Les concepts scientifiques* came out, and would die the following year. However, somebody like him, who regarded women's inferiority as a biological and evolutionary fact, was perhaps unlikely to take a great interest in a book authored by a woman.

11  Meyerson regarded himself as a Polish Jew, born in the 'old Russian empire' in 1859; he studied chemistry in Germany and then moved to France. In Paris, he divided his time between writing his philosophical books and the directorships of the Jewish Colonization Association and of the Rothschild Colonies in Palestine (Bensaude-Vincent and Telkès-Klein, 2016); (Meyerson, 1921, Meyerson, 1925, Meyerson, 1930 [1908]).

12  To give an idea of their connections, Meyerson, who at the time was unwell, wrote in a letter that Lévy-Bruhl visited him weekly and Brunschvicg almost as often (Meyerson, 1926). A few years later, writing to Lévy-Bruhl who was in Tunisia, Meyerson lamented that he felt lonesome at the times when Lévy-Bruhl usually visited him (when he was not away), despite the other visitors he had (Meyerson, 2009, Meyerson to Lévy-Bruhl, Letter of 25.6.1930, p. 415). Meyerson also kept the many notes that Brunschvicg, as well as his wife, sent to invite him to dinner (Meyerson, 2009, p. 83).

13  At the same time, she followed Meyerson's advice to submit her chapter on Stahl for publication in *Isis* as an article. Sarton clearly thought it was a good idea, as her article appeared in *Isis* in 1927 (Metzger, 1927c).

14  Meyerson was poorly, indeed he died that year of a heart attack, and almost certainly did not reply to her letter. A few months later, Metzger wrote him a card wishing him a happy (Jewish) New Year, and a prompt recovery, and expressing the hope that he had not forgotten her (Meyerson, 2009, p. 516).

15  It appears that they met in person for the first time in 1925, when Metzger invited the Sartons to dinner at her Parisian flat (Metzger, 1925b, Letter to Sarton of 17.7.1925).

16  (Metzger, 1922a); (Metzger, 1923b, Metzger, 1923a, Metzger, 1925a); (Metzger, 1926b, Metzger, 1927b, Metzger, 1927d); (Metzger, 1927a, Metzger, 1928b, Metzger, 1928c, Metzger, 1928a, Metzger, 1929, Metzger, 1931, Metzger, 1932a, Metzger, 1932b, Metzger, 1932c, Metzger, 1933a, Metzger, 1933b).

17  Berr advised Sarton on journal editorship (Berr, 1912, Letter to Sarton of 30.9.1912, Berr, 1913, Letter to Sarton of 24.5.1913), as in 1900 he had created the *Revue de synthèse historique* (*Revue de synthèse* from 1931).

18  Metzger obtained six votes; Léon Brunschvicg, full professor at the Sorbonne and one of the most prominent philosophers of the time, only two.

19  See her letters to the Singers (Metzger, 1933c, Letter of 4.6.1933); (Metzger, 1933d, Letter of 15.6.1933), and Singer's telegram to the International Committee offering his resignation, at Wellcome Institute Library, London, PP/CJS.

20  (Metzger, 1934, Metzger, 1935a, Metzger, 1936, Metzger, 1937–1939); these articles have been reprinted in (Metzger, 1987), with the exception of (Metzger, 1934).

21  Mieli had informed Sarton of his plans, perhaps because the latter was in the United States, which must have seemed far away from European trouble.

Mieli's letters to Sarton in the 1920s and 1930s appear to be mainly about the Academy's business, with the very notable exceptions of the 1927 letter in which Mieli explained what he thought of the Fascist regime and his reasons to move to France (Mieli, 1927), and of at least one in the late 1930s informing Sarton of his new move to Argentina (Mieli, 1939, to Sarton, Letter of 18.5.1939). His letters from Argentina in the 1940s are very different in tone from those from France: they are personal, and rather focussed on his health and isolation (Mieli, 1941–1949); (Chimisso, 2011).

22 Sarton wrote reference letters for Koyré, when the latter applied for an American visa. They also corresponded regarding Koyré's brother, although Sarton in November 1942 wrote to Koyré that it was too late for anybody to obtain visas to the United States (Koyré to Sarton 18.10.1941; 3.11.1942; 25.11.1942; Sarton to Koyré 6.11.1942).

## References

Académie Internationale d'histoire des sciences. 1935. VII<sup>e</sup> Réunion annuel. *Archeion*, 17, 227–258.

Anonymous. 1934. Nécrologie. Mort de Mme Pierre Curie. *Annales de l'Université de Paris*, 9, 384.

Anonymous. 1936. III<sup>e</sup> Congrès International d'Histoire des Sciences tenu au Portugal du 30 Septembre au 6 Octobre 1934: Actes, conférences et communications. Lisboa.

Bensaude-Vincent, B. & Telkès-Klein, E. 2016. *Les identités multiples d'Émile Meyerson*, Paris, Honoré Champion.

Berr, H. 1911. *La synthèse en histoire; essai critique et théorique*, Paris, Alcan.

Berr, H. 1912. Letters to Sarton. *Sarton papers bMs Am 1803 (1032)*, Cambridge, MA, Houghton Library.

Berr, H. 1913. Letters to Sarton. *Sarton papers bMs Am 1803 (1032)*, Cambridge, MA, Houghton Library

*Bollettino delle Ricerche*. 1930. 6/1/1930, 3/7/1930. *Fascicolo del casellario giudiziario*, 'Aldo Mieli', Rome, Archivio centrale di Stato.

Bourdieu, P. 1988 [1984]. *Homo academicus*, Cambridge, Polity Press.

Bourdieu, P. 1998 [1989]. *State nobility: Élite schools in the field of power*, Cambridge, Polity.

Bruhl-Metzger, H. 1914. Une théorie de la double réfraction chez Buffon. *Bulletin de la Société française de Minéralogie*, 37, 162–176.

Brunschvicg, L. 1912. *Les étapes de la philosophie mathématique*, Paris, Alcan.

Brunschvicg, L. 1922. *L'expérience humaine et la causalité physique*, Paris, Alcan.

Charle, C. 1984. Le beau marriage d'Émile Durkheim. *Actes de la recherche en sciences sociales*, 55, 45–49.

Charle, C. 1986. *Les professeurs de la Faculté des Lettres de Paris. Dictionnaire biographique 1909–1939*, Paris, Institut national de recherche pédagogique/Éditions du CNRS.

Chimisso, C. 2001. Hélène Metzger: The history of science between the study of mentalities and total history. *Studies in History and Philosophy of Science*, 32A, 203–241.

Chimisso, C. 2011. Fleeing dictatorship: Socialism, sexuality and the history of science in the life of Aldo Mieli. *History Workshop Journal*, 72, 30–51.

Chimisso, C. & Freudenthal, G. 2003. A mind of her own: Hélène Metzger to Émile Meyerson, 1933. *Isis*, 94, 477–491.

Clark, F. I. 1937. *The position of women in contemporary France*, London, P. S. King & Son.

Comité international et Centre international d'histoire des sciences. 1928. Membres du Comité international d'histoire des sciences: membres effectifs. *Archeion*, 9, 507–508.

Corsini, A. 1950. Aldo Mieli. *Rivista di storia delle scienze mediche e naturali*, 41, 111–113.

Delorme, S. 1947–48. Léon Brunschvicg. *Archives Internationales d'Histoire des Sciences*, 37, 519–521.

Delorme, S. 1983. Letter to Gad Freudenthal, 8.5.1983. Personal copy sent to me by Gad Freudenthal.

Diatta, C., Chef du Bureau de la Documentation Secretariat d'État. 1983. Letter to Gad Freudenthal, 15 Sept. 1983. Personal copy sent to me by Gad Freudenthal.

Dry, S. 2003. *Curie*, London, Haus.

Freudenthal, G. (ed.). 1990a. *Études sur / Studies on Hélène Metzger*, Leiden, Brill.

Freudenthal, G. 1990b. Hélène Metzger: Éléments de biographie. *In*: Freudenthal, G. (ed.) *Études sur / Studies on Hélène Metzger*, Leiden, Brill.

Gemelli, G. 1987. Communauté intellectuelle et stratègies institutionnelles. Henri Berr et la fondation du Centre International de Synthèse. *Revue de synthèse*, IV series, vol. 2, 225–259.

Guigue, A. 1935. *La Faculté des Lettres de l'Université de Paris depuis sa fondation (17 mars 1808) jusqu'au 1er Janvier 1935*, Paris, Alcan.

Koyré, A. 1986. *De la mystique à la science. Cours, conférences et documents 1922–1962, édité par Pietro Redondi*, Paris, Ed. de l'Ecole des hautes études.

Kuhn, T. S. 1977. *The essential tension: Selected studies in scientific tradition and change*, Chicago, University of Chicago Press.

Kuhn, T. S. 1996 [1962]. *The structure of scientific revolutions*, Chicago, The University of Chicago Press.

Lalande, A. 1899. *La dissolution opposée à l'évolution dans les sciences physiques et morales*, Paris.

Lalande, A. 1999 [1926]. *Vocabulaire technique et critique de la philosophie*, Paris, Presses Universitaires de France.

Metzger, H. 1918. *La genèse de la science des cristaux*, Paris, Alcan.

Metzger, H. 1922a. Émile Meyerson, *De l'explication dans les scienc*es, 2 vols. Paris, 1921. *Isis*, 4, 382–385.

Metzger, H. 1922b. L'évolution du regne metallique d'après les Alchimistes du XVIIe siècle. *Isis*, 4, 466–482.

Metzger, H. 1923a. André Cresson: Les réactions intellectuelles élémentaires. *Isis*, 5, 473–474.

Metzger, H. 1923b. Léon Brunschvicg, *L'expérience humaine et la causalité physique. Isis*, 5, 479–483.

Metzger, H. 1924. Letters to Sarton. *Sarton papers bMs Am 1803 (1032)*, Cambridge, MA, Houghton Library.

Metzger, H. 1925a. Émile Meyerson. *La déduction relativiste*. Paris, Payot, 1925. *Isis*, 7, 517–520.

Metzger, H. 1925b. Letters to Sarton. *Sarton papers bMS Am 1803 (1032)*, Cambridge, MA, Houghton Library.

Metzger, H. 1926a. La philosophie de la matière chez Stahl et ses disciples. *Isis*, 8, 427–464.

Metzger, H. 1926b. *Le genie de Pascal* by Léon Brunschvicg. *Isis*, 8, 175–176.

Metzger, H. 1926c. *Les concepts scientifiques*, Paris, Alcan.

Metzger, H. 1927a. André Cresson, *Les courants de la pensée philosophique française. Isis*, 9, 489–490.

Metzger, H. 1927b. Émile Meyerson. *Identité et réalité. Isis*, 9, 470–472.

Metzger, H. 1927c. La théorie de la composition des sels et la théorie de la combustion d'après Stahl et ses disciples *Isis*, 9, 294–325.

Metzger, H. 1927d. Lucien Lévy-Bruhl, *L'âme primitive. Isis*, 9, 482–486.

Metzger, H. 1928a. *Le progrès de la conscience dans la philosophie occidentale* by Léon Brunschvicg *Isis*, 10, 98–102.

Metzger, H. 1928b. *Les historiens de l'esprit humain* by Raymond Lenoir. *Isis*, 10, 501–502.

Metzger, H. 1928c. *Une nouvelle philosophie des sciences – le causalisme de M. Émile Meyerson* by André Metz. *Isis*, 11, 149–150.

Metzger, H. 1929. *Les fonctions mentales chez les sociétés inférieures* by Lucien Lévy Bruhl; *Les non-civilisés et nous, différence irréductible ou identité foncière* by Raoul Allier; *La raison primitive* by Olivier Leroy. *Isis*, 12, 343–347.

Metzger, H. 1930. *La chimie*, Paris, Boccard.

Metzger, H. 1931. *Maupertuis* by Pierre Brunet. *Isis*, 15, 177–179.

Metzger, H. 1932a. Émile Meyerson, *Du cheminement de la pensée*, 3 vols. Paris 1931. *Isis*, 17, 444–445.

Metzger, H. 1932b. *L'introduction des théories de Newton en France au XVIIIe siècle (Avant 1738)* by Pierre Brunet. *Isis*, 17, 433–435.

Metzger, H. 1932c. *Le surnaturel et la nature dans la mentalité primitive* by Lucien Lévy-Bruhl; *La mentalité primitive. Isis*, 17, 450–453.

Metzger, H. 1932d. Letters to Sarton. *Sarton papers bMs Am 1803 (1032)*, Cambridge, MA, Houghton Library.

Metzger, H. 1933a. *La philosophie de Fontenelle ou le sourire de la raison* by J. R. Carré. *Isis*, 19, 205–207.

Metzger, H. 1933b. *Le pluralisme cohérent de la chimie moderne* by Gaston Bachelard. *Isis*, 19, 233–235.

Metzger, H. 1933c. Letter to Dorothea Waley Singer. Wellcome Institute Library, London, PP/CJS.

Metzger, H. 1933d. Letters to Charles and Dorothea Singer. Wellcome Institute Library, London, PP/CJS.

Metzger, H. 1934. La philosophie de la matière chez les chimistes du 17ᵉ et du 18ᵉ siècles. *Thalès*, 1, 59–64.

Metzger, H. 1935a. La littérature chimique française aux 17ᵉ et 18ᵉ siècles. *Thalès*, 2, 162–165.

Metzger, H. 1935b. *La philosophie de la matière chez Lavoisier*, Paris, Hermann.

Metzger, H. 1936. L'évolution de l'esprit scientifique en chimie de Lémery à Lavoisier. *Thalès*, 3, 107–113.

Metzger, H. 1937–1939. Le rôle des précurseurs dans l'évolution de la science. *Thalès*, 4, 199–209.

Metzger, H. 1938a. Alchimie. Communication pour servir au Vocabulaire historique. *Revue de synthèse*, 18, 43–53.

Metzger, H. 1938b. *Attraction universelle et religion naturelle chez quelques commentateurs anglais de Newton*, Paris, Hermann.

Metzger, H. 1947a. Atome. Projet d'article pour un vocabulaire historique. *Revue d'histoire des sciences et de leurs applications*, 1, 51–62.

Metzger, H. 1947b. La science, l'appel de la religion et la volonté humaine. *Revue philosophique*, 137, 401–415.

Metzger, H. 1954. *La science, l'appel de la religion et la volonté humaine*, Paris, Boccard.

Metzger, H. 1969 [1923]. *Les doctrines chimiques en France du début du XVII$^e$ à la fin du XVIII$^e$ siècle*, Paris, Blanchard.

Metzger, H. 1974 [1930]. *Newton, Stahl, Boerhaave et la doctrine chimique*, Paris, Blanchard.

Metzger, H. 1987. *La méthode philosophique en histoire des sciences. Textes 1914–1939, réunis par Gad Freudenthal*, Paris, Fayard.

Metzger, H. 1987 [1929]. La philosophie de Émile Meyerson et l'histoire des sciences. *In:* Metzger, H. 1987. *La méthode philosophique en histoire des sciences. Textes 1914–1939, réunis par Gad Freudenthal*, Paris, Fayard.

Metzger, H. 1991 [1930]. *Chemistry*, West Cornwall, CT, Lucust Hill Press.

Metzger, P. 1913. *Le conseil supérieur et le grand bailliage de Lyon*, Lyon.

Meyerson, É. 1921. *De l'explication dans les sciences*, Paris, Payot.

Meyerson, É. 1925. *La déduction relativiste*, Paris, Payot.

Meyerson, É. 1926. [Letter to Høffding of 4 August 1926]. *In:* Brandt, F., Høffding, H. & Adigard Des Gautries, J. (eds.) *Correspondance entre Harald Høffding et Emile Meyerson*, Copenhagen, Einar Munskgaard.

Meyerson, É. 1930 [1908]. *Identity and reality*, London, New York, Allen & Unwind, MacMillan.

Meyerson, É. 2009. *Lettres françaises. Éditées par Bernadette Bensaude-Vincent et Eva Telkes-Klein*, Paris, CNRS.

Mieli, A. 1927. Letters to Sarton. *Sarton papers bMS Am 1803 (1039)*, Cambridge, MA, Houghton Library.

Mieli, A. 1928. Letters to Sarton. *Sarton papers bMS Am 1803 (1039)*, Cambridge, MA, Houghton Library

Mieli, A. 1930. La section d'Histoire des Sciences du Centre international de synthèse. *Bulletin de la Société Française d'Histoire de la Médicine*, 24, 122–126.

Mieli, A. 1932. La création du Comité International d'Histoire des Sciences et son activité actuelle. *Archeion*, 14, 357–358.

Mieli, A. 1939. Letters to Sarton. *Sarton papers bMs Am 1803 (1039)*, Cambridge, MA, Houghton Library.

Mieli, A. 1941–1949. Letters to Sarton. *Sarton papers bMs Am 1803 (1039)*, Cambridge, MA, Houghton Library.

Pomata, G. 2013. Amateurs by Choice: Women and the Pursuit of Independent Scholarship in 20th Century Historical Writing. *Centaurus*, 55, 196–219.

Rey, A. 1936. Institut d'Histoire des sciences et des techniques: Rapport du Directeur sur son organisation et son activité. *Annales de l'Université de Paris*, 11, 340–345.

Sarton, G. 1927–1947. *Introduction to the history of science*, Baltimore, William and Wilkins.

Sarton, G. 1942. Letters to Mieli. *Sarton Papers bMS Am 1803.1 (413)*, Cambridge, MA, Houghton Library.

Section d'histoire des sciences. 1930. Séance du 26 février 1930. *Archeion*, 12, 58–59.

Singer, C. 1946. Mme H. Metzger-Brül. *Nature*, 157, 472.

# Part I
# Themes in Metzger's writings

# 1  Anachronism and scientific concepts

I would like this thought in its nascent phase to be reconstituted in the mind of the reader, who must for a brief time be the disciple and contemporary of the past scholar, who confides in him through my pen

(Metzger, 'The philosophical method in the history of science')

## Reading past texts

When she wrote her first monograph, *La genèse de la science des cristaux*, Hélène Metzger had a degree in crystallography and no formal university training in history or philosophy. She chose to study sixteenth- and seventeenth-century texts that dealt with crystals, from a period in which, as she wrote, crystallography was not yet an independent science (Metzger, 1918, p. 5). The texts that she studied were remarkably different from anything in current science. How should they be read? How could she make sense of them? Discarding the option suggested by her former university teacher, namely to ignore them altogether and do something else, she had two general options in front of her. The first was to make use of her knowledge of crystallography and search past texts for similarities with current theories and suggestions of future discoveries. This option would have required her to be very selective with the authors whom she chose to include in her study. They would have been those consecrated by current science as 'fathers' of the discipline and as visionary 'great men' of science. In other words, she could choose to write a present-centred history. However, she discarded this option very decisively. Both her book on crystallography and her subsequent works on the history of chemistry were consciously constructed in opposition to proleptic narratives. She pledged not to judge past texts by the standards of modern science and argued that historians who did so constructed 'triumphalist' narratives out of a self-serving choice of superficially interpreted events.[1] She even mocked those historians of science: according to them, she wrote, scientists only need to reject absurd old prejudices and start experimenting without any preliminary idea in order to come up with 'good and solid' chemistry (Metzger, 1974 [1930], pp. 6–7). She chose the second general option, namely to read past texts in their historical specificity. More precisely, her objective was to read them as one of their contemporaries would have done. As a historian, she aimed to make herself a 'contemporary of the scholar whose work' she studied (Metzger, 2019 [1937]).[2] However, this was no easy task. Her sources often appeared to be not just a little odd, but in fact to be illogical, and at times simply incomprehensible. How to make sense of such assertions as those of the sixteenth-century medic and alchemist Oswald Croll (or Crollius), who suggested that the colour of flowers would reveal to us the colour of the stars? Or again that there must be as many species of wood in the world as there are bones in the human body (Metzger, 2019 [1936])? In the same period, Cardan and Biringuccio held that the calcination of a metal is like the death of a living being; in both cases the 'life principle', which makes beings light, escapes them (Metzger, 1969 [1923], p. 375). How to explain their lack of hesitation in comparing living beings and metals?

It was not just a matter of odd claims and methods. She was aware that the authors of the texts she studied had aims that were not only different from those of modern scientists, but also diverse. In her period of choice, neither crystallography nor chemistry was an established discipline, with a set of dominant theories, institutions and training programmes. Theories and practices varied widely, and so did professional profiles; in fact, many individuals who contributed to the formation of crystallography were interested only in collecting rare and beautiful objects in their cabinets of natural history (Metzger, 1918, p. 35). Some authors of her sources had no ambition to formulate any theory but nevertheless made significant discoveries, such as the double refraction in crystals of Iceland spar (Metzger, 1918, pp. 35–36). She faced analogous complexity in her histories of chemistry. She followed Herman Boerhaave (1668–1738) and his contemporaries, and presented chemistry as the fusion of two different traditions: that of the 'metallurgists' or alchemists, who aimed to transmute corruptible metals into gold, and that of the pharmacists who searched for a panacea for all diseases (Metzger, 1969 [1923], pp. 22–23). Those two traditions were not internally coherent, let alone fully fledged disciplines. She warned her readers that early modern sources should not be read as expositions of properly formed theories, or as relying on well-formed methods. For instance, she presented Wallerius' treatise on mineralogy (Wallerius, 1753) as a groundbreaking work that enjoyed a well-deserved success throughout Europe. At the same time, she pointed out that Wallerius' treatise is not guided by any particular method, nor does it afford a systematic exposition of the material, as a modern scientist would expect (Metzger, 1918, p. 58). Moreover, especially for the earliest times she studied, a great variety of practices and ideas co-existed: no theory could prevail or at least attract enough practitioners to create a coherent research programme. Metzger knew that she was not saying anything new when she pointed out the fragmented reality of past activities and theories concerning matter. She quoted the famous words of Fontenelle before the Académie des sciences to the effect that the spirit of chemistry was confused and concealed, and she mentioned the rather dismissive article dedicated to chemistry in the *Encyclopédie* (Metzger, 1991 [1930], p. 5). However, unlike those sources, she did not belittle early chemists' achievements. Rather, she simply saw early chemical writings as a challenge to the historian. She was determined to reconstruct a history of chemistry that, while not hiding its internal variety and fragmentation, did not dismiss early practitioners' efforts, or minimise their significance.

Metzger's sources were generally textbooks, presentations of chemical theories and descriptions of experiments. She intended to use them as gateways to the live and concrete activities of past practitioners. However, those texts not only revealed, but also hid, past practices. Before the publication of Lémery's *Cours de chimie* (Lémery, 1680 [1675]), pharmacists and medics avoided describing their procedures in full, in order to guard the secrets of their 'lucrative business' (Metzger, 1991 [1930], p. 31). Even when procedures were comprehensively described, many difficulties remained. As she remarked, the chemistry that was taught, and written for the educated public, was not the same as chemistry in the making. Her ambition was to visit, in a manner of speaking, the laboratories of the past, carry out the same experiments as those of early chemists, and share their aims and concerns (Metzger, 1969 [1923], p. 343). She wanted to understand past scholars as their students would have done, but her only sources were written texts. As she put it, Lémery's students could go into his laboratory and observe his demonstrations (Metzger, 1969 [1923], p. 322). By contrast, the historian is left wondering what he really meant when he named reagents and described procedures (Metzger, 2019 [1936]). She acknowledged several obstacles to a proper understanding of chemical procedures described in her sources. She hinted at something that later was called tacit knowledge:[3] Lémery's students were there to see what did not, in fact could not, end up in a report. There was an even more fundamental issue: how can the historian ascertain that the experiments described were really performed, rather than just dreamt up (Metzger, 2019 [1936])? These were difficult problems. On the one hand, Metzger thought that historians made mistakes that could in fact be avoided if only more care were taken. On the other hand, she thought that in order to understand past sources as well as we can, a great deal of work is needed. Both simple and complex misunderstandings of past texts for her generally originate from the projection of current knowledge onto the past. In other words, they are instances of anachronism. She did not use the term, but her analysis of anachronism is very detailed, though scattered throughout her books, as a sort of constant warning. She discussed many types of anachronism, which I have systematised into the following types. The most basic type, which I shall call terminological anachronism, stems from the assumption that our current terminology applies to past times, and that past practitioners employed a consistent terminology. A second type can be named material anachronism; this occurs when the historian assumes that past practitioners had access to the same substances as modern scientists do. Epistemological anachronism, in my terminology, relates to the formation of the scientific object and to epistemic values such as precision; taxonomic anachronism originates from the historian's failure to grasp the differences in classification and, indeed, the different degree of importance attached to classification in different epochs. Finally, unless historians understand past scholars' worldviews, they would commit what I call metaphysical anachronism. I shall analyse those types of anachronism in Metzger's writing in the next section.

Metzger aimed to visit, albeit virtually, not only past chemists' laboratories, but also their minds, to share their assumptions and views of nature. In line with the concerns of philosophers and ethnologists of her time – as we shall see in Chapters 5 and 6 – she aimed to capture the ways of thinking of a time and place, rather than the mind of an individual chemist. In order to understand past sources, for her, the most important and complex step that historians should take is to grasp the concepts that past scholars employed to organise their empirical data, and indeed their world. My discussion of her analysis of those concepts will follow that of anachronism.

## Varieties of anachronism

The type of natural philosophy that Metzger studied was first of all manipulation of matter, and, as mentioned above, some early practitioners were not even interested in theory. Many others were interested in elaborating theories, and these were the chemists who secured prominent positions in the history of chemistry. Nevertheless, as she pointed out, the aim of chemical theory is to explain material reactions observed in the laboratory (Metzger, 1974 [1930]). The centrality of material objects is beyond doubt. It is not by chance that the first substantial chapter of Metzger's popular history of chemistry is dedicated to a description of the seventeenth-century laboratory. She opens this chapter with the following words:

> In order to judge correctly the effort of chemists in times past, we must first of all be aware of their working methods. We must know with what kinds of reagents they experimented and be acquainted with the types of tools they used when modifying them. In short, we need to visit one of their laboratories.
>
> (Metzger, 1991 [1930], p. 6)

Material objects present the historian with the most obvious, but in fact quite insidious, form of anachronism. At the most basic level, there is the possibility of error due to terminological misinterpretation, as the language that past scholars used was different from modern usage. More problematically, their terminology was not standardised; indeed, they employed everyday words, which were subject to variations. Metzger admitted that the historian cannot but use conjectures in order to establish to which substances her sources refer (Metzger, 1969 [1923], p. 317). However, she also pointed out that some historians slipped into error by assuming that the same name, e.g. 'sulphur', indicated the same element as it does nowadays.[4] In fact, this could not be the case as early chemists claimed to have analysed and synthetised sulphur (Metzger, 1969 [1923], p. 403). Analogously, in Nicolas Lémery's chemistry handbook, 'arsenic' did not indicate the element, but rather an arsenic sulphide that can be found in nature (Metzger, 1969 [1923], p. 318).[5] This is not just a terminological issue. The substances that seventeenth-century chemists studied differed in many ways from those of the modern chemist. Notably, they were not pure substances of which a modern chemist could determine the formula (Metzger, 1991 [1930], p. 10), as early chemists would not have had the technical means to obtain pure substances, even if they had wanted to do so. They also did not have the means to work on identical substances over time and across places, and for this reason it was often difficult to get consistent results. Although all this may sound obvious, Metzger believed that some historians did not pay enough attention to the difference between our material world (not her words) and that of her early practitioners. The tools that the latter used were imprecise, from a modern point of view. For instance, one of the most important tools was fire, so much so that these practitioners were often called 'the philosophers of fire' (Metzger, 1991 [1930], p. 9). It was however difficult for them to keep their fires at a constant temperature, or even to ensure that their containers would not break (Metzger, 1991 [1930], p. 11). In short, Metzger aimed to bring to her readers' attention that the material conditions under which early chemists worked, as well as the objects of their investigation, were different from those of a modern chemist.

Terminological and material anachronism point to a more complex type of anachronism, which we can call epistemological, and that concerns, among other things, how the scientific object is constructed. It is not only the case that early chemists could not have purified substances to the degree that is possible nowadays, but also that they did not aim to do so. In Metzger's presentations, early crystallographers, chemists, medics, pharmacists and alchemists appear to aim to study natural objects, an aspiration that seems in fact quite reasonable. That the study of nature should construct objects that differ considerably from those of everyday experience may look obvious to people acquainted with modern science, but it may seem rather strange otherwise.[6] Precision was not always pursued: she pointed out that, up to Lavoisier's times, the balance was listed as an optional tool in the laboratory (Metzger, 1991 [1930], p. 7). What was held as important to measure also varied, although these variations may not be captured in terminology. For instance, different scholars referred to the measurement of 'quantity of matter'. Van Helmont (1579/1580–1644) and Ole Borch, or Borrichius (1626–1690), employed a balance to measure it. By contrast, Cartesians would have no longer done so, as they identified it with extension, or volume (Metzger, 1969 [1923], p. 387). Metzger concluded that it is an illusion for the historian to hope to offer precise definitions of the concepts of chemistry, as their meanings have slowly changed (Metzger, 1974 [1930], p. 11). The historian faces similar difficulties with classifications of natural objects. To a modern person, classification in crystallography is crucial. However, Metzger pointed out that collectors of stones may not have been necessarily interested in classifying them. Some scholars did, but their classifications did not play the same role as in later research. She cited the example of Wallerius, who presented a classification of minerals according to their chemical properties but did not seem to attach any special importance to it (Metzger, 1918, p. 59). Before even classifying crystals, one may expect a definition of the objects of this science, i.e. crystals. However, Metzger's ambition was to study the formation of this science, and this included reading texts that did not have any such definition, but rather presented descriptions of 'angular stones'. To name a couple of authors that Metzger considered at length, neither Nicholas Steno (1638–1686) nor La Hire in his article on Iceland spar (1710)[7] defined crystals. Unsurprisingly, they also failed to classify them according to either their formation or their differences in composition (Metzger, 1918, p. 43). Moreover, scholars had to face views expressed by authorities such as Buffon, who denied that crystallisation was an important property at all (Metzger, 1918, p. 70). In *La genèse*, Metzger described the first attempts at classifying crystals, which were diverse not only in their comprehensiveness, but also in their criteria. In general, those early attempts for her were completely 'artificial and arbitrary'. For instance, dictionaries of chemistry and natural history proposed an alphabetic order to nature, which was soon found unsatisfactory; other tentative classifications were proposed, used and later discarded (Metzger, 1918, p. 221). What was important for her is that the historian neither dismisses these early attempts nor projects her modern understanding of crystallography onto them. In other words, the historian should not commit a type of anachronism that we could call taxonomic anachronism. Metzger connected the lack of classifications, and, later on, of universally accepted classifications of crystals, to the more general issue of the sometimes astonishingly different ways in which past scholars classified their objects. Metzger illustrated this issue in a chapter of *La genèse*, in which she described how crystallography gained autonomy from the study of living beings. Crystals, which can nowadays be jokingly described as 'among the "deadest" objects in the universe' (Glazer, 2016, p. xiii), were often seen as living beings, not least because of their growth (Metzger, 1918, p. 17). Among her many examples is that of the botanist Joseph Pitton de Tournefort (1656–1708), who, as Metzger put it, saw plants everywhere: he considered corals and sponges as plants, and then extended his theory of growth and development of corals to stones (Metzger, 1918, pp. 94–95). Another example is that of Jean-Baptiste Robinet (1735–1820), who argued that 'angular stones' reproduced by seed, and saw in fossils like shells and bones small mouths from which minerals supposedly got their nourishment (Metzger, 1918, p. 121).

The blurred boundaries between what we now consider minerals, plants and animals point to a fundamental issue for the historians. Past scholars appear to have a view of the natural world that greatly differs from the modern view (see Gibson, 2015). They also seem to make inferences from one case to another that look completely unwarranted to the modern reader. Metzger was particularly keen to be faithful to past scholars' worldviews, and to uncover the underlying assumptions that made possible for them to explain phenomena in those particular ways. She was determined to avoid what we could call metaphysical anachronism. It was not just a matter of the gulf between the modern historian's and the early chemist's metaphysics. In the culturally fragmented times that she studied, one scholar could declare another difficult to understand, or indeed dismiss another's theory as just reverie. William Davidson's *Philosophia Pyrotechnica* (1640)[8] for Metzger is not only difficult for modern people, but it would have been hardly readable for a Cartesian, who would not have shared the basic assumption that the philosopher must go behind the objects of our senses, as they are only copies, or copies of copies, of the real things (Metzger, 1969 [1923], pp. 45ff.). The modern historian is faced with even greater challenges: she admitted that her first reaction to Du Clos'[9] pamphlet against Boyle's theory resembled that of Fontenelle's to chemistry, mentioned above (Metzger, 1969 [1923], p. 269). However, without naming it, Metzger believed that readers who dismiss admittedly difficult works commit metaphysical anachronism. They applied their own metaphysics to those works, which, as a consequence, became opaque to them. By contrast, she set out to understand their metaphysics, worldviews, assumptions and aims, accepting that they could be rather different from those of modern people, as well as from those of scholars who lived in other periods. In short, her principal aim was to understand what she called authors' 'mentalities' following the terminology and scholarship of her time. As I shall briefly note in this chapter, she acknowledged the important role that Lucien Lévy-Bruhl's concept of mentality played in her own reflection on the human mind and her concept. In Chapter 6, I shall discuss more fully her critical evaluation of Lévy-Bruhl's view of the human mind. She opted for a study of what she called the '*a prioris*' of human thought, to mean all the components of knowledge that do not come directly from experience, but that order and shape that experience. Her view of mental *a prioris* is very complex, including as it does concepts as well as a-rational elements, such as emotions and habits. I shall discuss the latter in Chapter 3. Here, I shall start from her analysis of the concepts that order, and have ordered in the past, human experience of nature. For her, knowledge of these concepts is necessary in order not to commit anachronism.

## Scientific concepts

Metzger observed that until the seventeenth century, pharmacists and medics had almost exclusively employed substances of animal and vegetable
origin. As she wrote, this was the case even though, after Paracelsus, mineral remedies were known, and they were known to have very different
properties from animal and vegetable products. Practitioners of a newly
constituted science, chemistry, in order to analyse their mixts, continued
to employ pharmaceutical practices, which they extended to mineral substances. They used fire to burn, distil and dry, water to dissolve, and acids
to corrode; they also made use of the same tools as previously, including
alembics, stoves and mortars (Metzger, 1969 [1923], p. 347). Her question
was the following: why did they think that their techniques and tools would
work, as a matter of course, on inorganic substances? Metzger provided
two answers to this question. The first is that the human mind exhibits a
natural tendency to think that what it does not still know is built on the
same model of what it already knows. The second answer is that the philosophy of the time regarded analogy as the key to unlocking the secrets of
nature (Metzger, 1969 [1923], p. 348). This double answer well illustrates
how Metzger interpreted the choices and views of the scholars she studied.
She constantly looked for general tendencies of the human mind, and at the
same time for ways of thinking and acting that belonged specifically to the
time under study. For her, the specificity of a way of thinking often rests on
the particular manner in which a concept, which is present in all times, is
being used. For instance, pharmacists used analogy. The concept of analogy was obviously not peculiar to them; however, the role that this concept
played in their way of thinking and in the way in which they saw the world
was very different from that of a modern person. Metzger aimed to uncover
what made past scholars conduct their thoughts and their experiments in
ways that may look utterly bizarre to the modern reader, but that would
have seemed perfectly plausible to their contemporaries.

Her concern with anachronism stimulated her to find out which concepts scholars employed, consciously and above all unconsciously, in the
organisation of their experiences. She aimed to find out the order these
scholars imposed on nature. Although the monograph that she entirely
dedicated to this issue is entitled *Les concepts scientifiques*, she was not
primarily interested in well-defined concepts of a mature science. Rather,
as she wrote, she wanted to capture that moment in which thought organises 'imprecise notions, vague claims, half-obscure theories, prejudices and
intuitions that are characteristic of the common sense of a certain age' into
a coherent system (Metzger, 1926, p. 1). Part of her book was inspired by
a recent work by André Cresson on 'elementary intellectual reactions' that
she had reviewed for *Isis* (Metzger, 1923). Cresson's essay is largely dedicated to analogy, which, according to the author, is employed instinctively
by human beings (Cresson, 1922).[10] Metzger's enterprise, however, differed

from Cresson's on a number of points. First of all, her sources were from the history of science: what interested her was to find out the mental mechanisms behind scientific theories and laboratory practices. Second, her work reached further, as she aimed at presenting a catalogue of concepts used in natural philosophy and science, including not only analogy, which she subdivided into three types, but also concepts founded on permanence of substances and types of evolution. She also dedicated a concluding chapter to the nominalist approach, which in her view differed from all the preceding ones in that it did not consider scientific concepts as means to learning 'something' about the objects to which they are applied (Metzger, 1926, p. 145). Cresson's book, nevertheless, lent philosophical support to her view that analogy is one of the most fundamental forms of inferences in human reasoning. Analogical reasoning for her was arguably the hallmark of what she called 'expansive thought'.[11] She counterposed expansive thought to reflective thought, and described the former as follows:

> By *expansive* thought, I mean thought which rushes forward tumultuously and simultaneously in all the directions through which it can fight its way, which advances constantly and irregularly without stopping to consider the ground covered, and without attempting to build a theoretical monument!
>
> (Metzger, 2019 [1936])

Expansive thought for her dominates 'science in the making', both as early formation of a discipline and as creative moments within an established science, as we shall see in the next chapter. Analogy, as she explained in her 'The *a priori* in scientific theory and the history of science' (Metzger, 2019 [1936]), plays a central role in expansive thought, and therefore in the early science that she studied, because for her human beings instinctively place the various objects of their experience in a relation of analogy. However, although for her early science is in closer continuity with our instinctive and everyday way of thinking, the link is never severed.

She called the most basic type of analogy 'virtual analogy'. By 'virtual', Metzger meant 'hypothetical'; this analogy is hypothetical mainly from the point of view of the historian, as many past scholars accepted it as 'real' or 'material' (Metzger, 2019 [1936]). Virtual analogy is at work when, from a simple observation of similarity among objects, we leap to the provisional and untested conclusion that these objects must also share other properties, or indeed be specimens of the same natural kind. By relying on analogy, Christiaan Huygens (1629–1695) argued that since the disposition of organs and tissues is analogous in all bodies, we can conclude that the other planets are just like Earth (Metzger, 1926, pp. 19–20). For Metzger, analogy is so fundamental to our thinking that even scholars whom we would not expect to use it did in fact employ it. She cited Descartes, who, despite his rejection of non-mechanistic explanations, suggested to Huygens that expansive diamonds should be substituted for cheap glass in the latter's experiments aimed at establishing whether a substance that dissolved gold would have the same effect on a precious stone (Metzger, 1926, p. 20). However, this basic type of analogy for Metzger has also produced important results. For her, it was virtual analogy that guided Boyle and Lavoisier to relate combustion and respiration (Metzger, 1926, p. 24).

The greatest asset of Renaissance natural philosophy was another type of analogy, which she named formal analogy. The whole worldview of that period was dominated by similarity in the forms of things, first of all the correspondence between macrocosm and microcosm. The seven so-called errant planets found their counterparts in the main organs the human body, with the sun being the heart, the moon the brain, Saturn the spleen, Mars the gall bladder, Venus the kidneys, Mercury the lungs and Jupiter the liver. Metzger explained that Paracelsus rejected astrological influences, and rather taught his disciples that heavens and human beings were two different worlds, between which there was a series of analogies (Metzger, 1926, p. 29). Most Paracelsians would have also accepted a correspondence between heavenly bodies and metals, with the sun and the moon, more luminous than any other, corresponding respectively to gold and silver, the most precious metals. They therefore recognised three orders of objects that stood in analogical relationships: heavenly bodies, organs and metals. This led to the correspondence between metals and organs, which linked gold with the heart, and silver with the brain (Metzger, 1926, pp. 29–30). Metzger saw this type of analogy at work in a number of other philosophies and systems, including Leibniz's pre-established harmony, and Descartes' mind-body dualism (Metzger, 1926, p. 31).

Given Metzger's focus on the history of chemistry, it is not surprising that the type of analogy that she detected most often was 'active analogy', in her terminology. Active analogy is the action of something on something to which it resembles. The 'action' of a substance over another was, needless to say, the central concern of chemists. They had a specific interest in finding substances that would react with others in the desired manner. Unlike formal analogy, which normally affects large systems, and typically the cosmos, active analogy works at all levels. Metzger believed that the three types of analogy occur in a natural development in the human mind: at first human beings notice that some things are similar, and they group them together (virtual analogy); then they put classes of objects in a sort of structural correspondence (formal analogy); finally, they believe that those classes of objects have properties of mutual attraction or rejection (active analogy). She detected this succession of types of analogies also within the thought of a single scholar. One of her examples is William Davidson. Consistently with his Neoplatonic metaphysics, as I mentioned above, he held that the sensible world is a copy of the immutable and perfect world that God created. It is thanks to formal analogy between the poor copy that we experience and the perfect original that we can gain genuine knowledge.

At the same time, Davidson relied on active analogy to explain chemical reactions: for him, one can successfully dissolve a substance into another because solvent and solute are similar. For instance, he claimed that there is an affinity between gold and aqua regia, and this is the reason why the latter can dissolve the former (Metzger, 1926, pp. 41–42). If there is no similitude between two substances, then one will not dissolve into the other. For Metzger, active analogy was often at the core of past assertions and theories that modern historians generally find either incomprehensible or absurd. How could it be that medics really thought that a plant would have a beneficial effect on a human organ of similar shape? And yet, lungwort (*pulmonaria*) would be recommended for afflictions of the lungs, as the shape of its leaves reminded one of the human organ of respiration. Similarly, toothwort (*dentaria*) would be recommended for tooth problems, because its roots look like teeth, as the eighteenth-century physician Robert James explained in his medical dictionary (Metzger, 1926, p. 40). However outlandish these beliefs seem to us, for Metzger they were not irrational. Indeed, they were instances of a dominant worldview, namely that similar objects in the world are in mutual relation of influence. For her active analogy, just as the other types of analogy, is a way in which the mind organises experiences, and it is at work in various times and places. She relied on the work of Lucien Lévy-Bruhl to show that active analogy is at the core of what he called primitive mentality, supposedly observed in indigenous people of Africa and Australia (Metzger, 1926, pp. 37–38). Indeed, she made a direct connection between active analogy and what Lucien Lévy-Bruhl called law of participation, with all due differences.[12] For her, this shows that active analogy is a fundamental category of our thought, and as such it is not limited to early texts in the history of science. In fact, she found active analogy at work across times. She argued that it was the fundamental concept that guided Georg Stahl's theory of affinities. Stahl and his school took for granted that some elements react and bond with one another because they stand in a relation of affinity, or analogy (Metzger, 1974 [1930], pp. 149ff). Similarly, it was active analogy that guided Newton to reject Cartesianism, and towards his law of universal gravitation. Cartesianism, with its purely mechanical view of the universe, would not allow action at a distance. Metzger argued that Newton employed the natural tendency of the human mind to suppose that 'like acts on like', that is active analogy. Needless to say, she did not suggest that active analogy provided Newton with his law, but rather that it enabled him to engage with the possibility of a force exerted between distant objects (Metzger, 1938, p. 56). For her, active analogy may be characteristic of expansive thought, but, as we shall see later in this book, its creativity could never be expelled from science without destroying its development at the same time.

Virtual analogy suggested that similar objects should have similar properties. However, chemists observed that bodies that were not identical had nevertheless the same properties. How to explain this? For Metzger, scholars adopted other concepts to organise those experiences that could not be satisfactorily explained by analogy, or by analogy alone. The solution to the puzzle of different bodies displaying the same properties was that they must contain the same substance. For instance, early chemists knew that sulphur is completely combustible; they inferred that all combustible bodies must contain sulphur, and that therefore they form a class of 'sulphurous bodies'.[13] The concept of substance enabled chemists to classify chemical bodies according to the substance that those bodies were supposed to contain, based on their properties. Different properties would show that these bodies were made up of different substances. For instance, Boerhaave claimed that metals must be considered as made up of two substances: one giving them their 'metallic' state, the other their 'specific' state, as for instance gold or silver state. This, of course, encouraged chemists in the doomed attempt to decompose metals into two different substances (Metzger, 1926, p. 56). The concept of substance also enabled chemists to solve the problem of similar bodies displaying different properties, in several ways. One way was to suppose that these bodies contained the same substance in different proportions. In this case, Metzger took her example from more recent chemistry: the element O, which can occur in the form of oxygen [O], and ozone [$O_3$]. Another solution is that the same substances are in a different state, as white and red phosphorus (Metzger, 1926, p. 65). On the other hand, some practitioners who had a rigid substantialist approach explained everything by the presence or absence of substances. Metzger cited seventeenth-century scholars who explained water's transformation into ice by the mixing of the substance of water with frigorific substance (Metzger, 1926, p. 70).

Metzger proposed that another crucial concept in the study of nature in general, and of the transformation of matter in particular, is that of evolution. The line that evolution follows has been traced in different manners. The first manner for her is that of 'parallel evolution'. Her example is that of an apple tree and a pear tree. In spring, they can both be observed undergoing parallel changes, in terms of production of new leaves, flowers, unripe fruits and finally ripe fruits. These two trees are different, and produce different fruits, but they evolve in a parallel manner. For her, the concept of parallel evolution has a double aspect: one as cyclical evolution (as in the fruit trees examples), and the other as teleological [*finaliste*] evolution, as in the case of the particular organism that follows a linear evolution up to its death. Teleological evolution for her was the concept that organised the experiences of seventeenth-century alchemists. For them, gold is the only perfect metal; the other six (lead, tin, iron, copper, mercury and silver) evolve towards it. One day, all metals will reach the end of their evolution and become gold (Metzger, 1926, p. 93). Indeed, it was believed that one of the reasons why metals other than gold can be mined is that men, in their avarice, extract these metals too early, before they complete their evolution. On the other hand, the alchemists' aim was indeed to accelerate the natural evolution of metals and turn them into gold. Each lump of metal will follow a similar evolution to all other lumps of metals, although at different times. Evolution can also be seen as 'divergent', when different species are formed from a common origin. This concept of evolution is at the core of Charles Darwin's theory of evolution, as well as Jean-Baptiste Lamarck's (Metzger, 1926, p. 125ff).[14] Metzger also discussed the mirror image of this type of evolution, namely 'convergent evolutions'. In order to illustrate the latter, she cited the same example that she had employed to explain parallel evolution, namely the evolution of metals towards a common state, that of gold (Metzger, 1926, p. 139). In other words, she interpreted the view of metals as evolving towards the perfection of gold with both the concept of parallel evolution and that of convergent evolution. This may seem a little strange. However, we should not interpret Metzger's systematisation of different concepts as suggesting that concepts are mutually exclusive, as obviously different concepts are used at the same time. For instance, for her the philosophers of metals who intended to transform 'imperfect' metals into gold employed both the concept of teleological evolution and the concept of analogy. Their view of metal perfection and perfectibility for her rested on an analogy with the perfectibility of living organisms: imperfect metals are in the same relation to gold as a young person is to an adult, and the unripe fruit is to a ripe fruit (Metzger, 1926, p. 93). In her words:

> One finds constantly in the writings of [the] adherents [to the philosophy of metals] that 'the imperfect metal is to the perfect metal what the child is to the adult, what the green fruit is to the ripe fruit'. A child grows, a fruit ripens; this is what suggests the idea of perfecting, the idea of progress, the idea of evolution.
>
> (Metzger, 2019 [1937])

Moreover, Metzger did not suggest that concepts are used exactly in the same manner, either synchronically, by different groups in different places, or diachronically across generations of practitioners. Different concepts employed together acquire or lose dominance. There is also a more complex point, which we can infer from Metzger's presentation of her historical material: the historian and the philosopher should assume neither that past scholars held a perfectly coherent worldview, nor that even the same scholar's practices and pronouncements were always consistent.

Metzger aimed to investigate concepts as employed in science in the making, in other words at a stage in which these concepts were often not better defined than those of the general public. In *Les concepts scientifiques*, she did mention concepts that are the results of scientific and philosophical research, such Charles Darwin's and Herbert Spencer's evolution, Bergson's 'creative evolution', and André Lalande's 'dissolution' (later renamed 'involution').[15] However, not only did she spend very little time on them, but she often referred to them only in order to show that they reflected concepts that had previously guided research into nature, or were already part of scientific classification. For instance, she almost dismissed the philosophical controversies about the concept of evolution. Philosophers might raise doubts on the type of evolution that Darwin and Lamarck presented, she argued, but science accepts it, and the scientific classification of plants and animals suggests it. For her, the work of early chemists and alchemists already showed that the concept of evolution guides the mind in its organisation of empirical data. It goes without saying that she did not think that all forms and applications of the various concepts of evolution played a positive role in the investigation of nature. Concepts of evolution guided Charles Darwin, and scientists in their study of radioactive elements, but they also led alchemists to their doomed quest to transform 'imperfect' metals into gold (Metzger, 1926, pp. 139–140). Similarly, virtual analogy guided Boyle and then Lavoisier to regard respiration and combustion as similar processes, but also promoted the idea that the study of organs in a body could inform our understanding of planets, as seen above. Metzger's analysis of scientific concepts aimed to show that past scholars' apparently outlandish beliefs and theories were not just a 'sweet folly' (Metzger, 2019 [1936]). In fact, for her those beliefs could be informed by the same concepts as solid scientific theories, although differently employed and combined. Her study of concepts also yielded another remarkable, though not openly argued, philosophical lesson. This is that the study of nature has been dominated sometimes by one or more concepts and sometimes by others, but the success of a theory or a system is not necessarily linked to the use of particular concepts. There is no set of concepts that has been proven successful in all cases. More controversially, she clearly suggested that there is no 'right' way of classifying nature; in fact, different ways may lead to discoveries and enable scholars to manipulate nature successfully. Did she support a version of relativism? In reality, Metzger was interested neither in answering that type of epistemological question, nor in judging the theories presented in her primary sources.

Rather than judging, her ultimate goal was to understand the minds of past scholars, not as individuals, but as a community and as part of a past culture. Her goal presented many problems: as we have seen, past scholars' language, concepts and worldviews were far removed from those of modern science. How can then the historian first understand these texts, and second explain them to her readership? For Metzger, the historian's task is indeed complex. However, she did not think it was impossible for a modern person to understand past ways of thinking; as we would put it nowadays, she did not think that there was a problem of incommensurability. She always analysed ways of thinking at two different levels: universal and local. A good example is her discussion, cited above, of seventeenth-century pharmacists' confidence that the techniques and tools that they had previously used on organic substances would work on inorganic substances. As mentioned, she explained their confidence by claiming that the human mind exhibits a natural tendency to think that what it does not yet know is built on the same model of what it already knows; she also pointed out that the philosophy of the time regarded analogy as the key to unlocking the secrets of nature (Metzger, 1969 [1923], p. 348). It goes without saying that the historian can follow arguments based on ways of thinking that belong to humanity, in any time and place. What about the local *a prioris*? These for Metzger can be strikingly different from our own. However, Metzger presented these local ways of thinking as variations and different uses of universal *a prioris*. A good example is what Metzger called Renaissance science, which as she often pointed out, appears utterly alien to a modern reader. This is mainly because in the Renaissance analogy was the foundation of an entire world-view, and the main interpretative tool in the study of nature. This is completely different from modern science, as nobody would think that the shape of a plant's leaf would have any bearing on its medicinal effects on an organ of a similar shape. However, analogy is found in all times and places, from the thought of so-called primitives that Lévy-Bruhl described in his work, to Stahl's theory of affinities and Newton's law of universal gravitation. Crucially, analogy is employed in everyday thought, and for her the historian would know well what it is. The historian's challenge is rather to detect analogy in domains in which a modern person would not use it, and to realise that in the past analogy could dominate the understanding of all natural phenomena. However, there is no insurmountable obstacle to comprehension: it is not the case that modern people do not have the concept of analogy or evolution; it is rather their applications that are profoundly different. The knowledge of the specific uses of 'analogies, similitudes and sympathies' makes Renaissance thought 'permeable' to modern minds, and makes the alchemists' hopes of transmuting metals understandable (Metzger, 1969 [1923], p. 289). For her, this historian's task is to explain the otherness of past sources by showing the different uses past scholars made of concepts that are familiar to modern people.

## Notes

1  She cited as a current example of this traditional narrative Maurice Delacre's works (Metzger, 1974 [1930], p.7, fn.1). Delacre was a chemist turned historian, author of a history of chemistry (Delacre, 1920).

2  Henk H. Kubbinga has characterised Metzger's study of Stahl as from the point of view 'of a past chemist who witness the emergence and development' of his theory (Kubbinga, 1990).

3  See Polanyi (1958), Chapter 4, Polanyi (1966), and more recently Collins (2010).

4  As an example of a historian who committed terminological anachronism, she mentioned Delacre (see note 1).

5  She referred to the French original of Lémery (1680 [1675]).

6  The construction of the scientific object as different from the natural object was an epistemological theme of which Metzger was aware; philosophers she came to know well, including Léon Brunschvicg and Gaston Bachelard, wrote about it. As we shall see later in this book, however, her take on the 'appropriate' objects of science differed from that of those philosophers, especially Bachelard.

7 Metzger often disregarded first names, but in the case of La Hire, she probably did not know whether the author was Philippe or his son Gabriel. The publication to which Metzger referred does not show his first name (La Hire, 1710). Doubt about the identity of La Hire was also expressed more than 50 years after Metzger's *La genèse* by Jed Z. Buchwald (1980).

8 William Davidson, or Davison, used the Latin version of his name in his major work: Davissonus (1640).

9 Du Clos, Duclos, or indeed Cotreau. Metzger must have known little of his biography, which decades after her death was still pronounced obscure; see Todériciu (1974).

10 Metzger knew Cresson, a teacher at prestigious Paris *lycées*. In his book about philosophy in French universities, Célestin Bouglé, director of the École normale supérieure, mentioned Cresson as one of the influential *lycée* teachers, and regarded his works as inspired by positivist and naturalist philosophy (Bouglé, 1938, pp. vii–viii). Cresson was a prolific writer of works on the history of philosophy, and was a friend of the philosopher Léon Brunschvicg. Metzger mentioned in a letter to Sarton that Cresson had sent her two volumes, which she offered to review for *Isis*, as she had a positive evaluation of them (Metzger, 1927b); her review of Cresson's *Les courants de la pensée philosophique française* (Cresson, 1927) was then published (Metzger, 1927a).

11 Elsewhere, she talked of 'spontaneous thought'. 'Expansive thought' and 'spontaneous thought' are closely related concepts. Metzger used the former generally when emphasising scientific creativity and discovery, and 'spontaneous thought' more in general to indicate the instinctive ways in which human beings organise sense data.

12 Metzger was always up to date with her uncle Lévy-Bruhl's writings, and in fact alerted Sarton of their imminent publication, and offered reviews; see Metzger's Letter to Sarton of 18.5.1922 (Metzger, 1922) (this letter is partly reproduced in Freudenthal, 1990, but the relevant sentences do not appear there), and Metzger, Letter to Sarton of 14.4.1927, in (Freudenthal, 1990, p. 255). At the time of *Les concepts scientifiques*, she had certainly read (Lévy-Bruhl, 1925 [1922]); and almost certainly (Lévy-Bruhl, 1910), among his other writings. She reviewed a number of his works: Metzger (1927c), Metzger (1929), Metzger (1932), Metzger (1935). She also specifically discussed his work in relation to the history of science: Metzger (1930), now in Metzger (1987).

13 Metzger emphasised that this way of looking at substances has been fundamental to the study of matter from Paracelsus up to Lavoisier, who established that combustibility depends on whether a body can combine with another substance contained in its environment (Metzger, 1926, p. 55).

14 Darwin's *The origin of the species* was so well known that Metzger did not cite it (Darwin, 2009 [1859]). She however quoted from Lamarck's *Zoological philosophy* (Lamarck, 1914 [1809]; Metzger, 1926).

15 (Metzger, 1926, pp. 137–143). Lalande fashioned his concept of involution in opposition to that of evolution, especially in Herbert Spencer's version, which in his view had penetrated French philosophy at all levels. Lalande claimed that rather than a progressive differentiation, if anything we could observe a progressive homogenisation, something that also resonated with his democratic and egalitarian ideas (Lalande, 1899; Lalande, 1947). Bergson's *Creative evolution*, just as his other works, was well known in France to any intellectual, and indeed the general public (Bergson, 1998 [1907]). Spencer's works were also well known, and Metzger did not cite either *The principles of ethics* or *The principles of sociology* (Spencer, 1900 [1879–93], Spencer, 2003 [1897]). She cited other modern works, including Anthony (1921) and Osborn (1918).

# References

Anthony, R. 1921. *Le determinisme et l'adaptation morphologique en biologie animale*, Paris, Doin.

Bergson, H. 1998 [1907]. *Creative evolution*, Mineola, New York, Dover.

Bouglé, C. 1938. *Les maîtres de la philosophie universitaire en France*, Paris, Maloine.

Buchwald, J. Z. 1980. Experimental investigations of double refraction from Huygens to Malus. *Archive for History of Exact Sciences*, 21, 311–373.

Collins, H. M. 2010. *Tacit and explicit knowledge*, Chicago, IL, London, University of Chicago Press.

Cresson, A. 1922. *Les réactions intellectuelles élémentaires*, Paris, Alcan.

Cresson, A. 1927. *Les courants de la pensée philosophique française*, Paris, Colin.

Darwin, C. 2009 [1859]. *On the origin of species: By means of natural selection, or, the preservation of favoured races in the struggle for life*, London, Penguin Classics.

Davissonus, W. 1640. *Philosophia pyrotechnica ... Seu Cursus chymiatricus, etc*, Paris, Apud Ioannem Bessin.

Delacre, M. 1920. *Histoire de la chimie*, Paris, Gauthier-Villars et cie.

Freudenthal, G. (ed.). 1990. *Études sur / Studies on Hélène Metzger*, Leiden, Brill.

Gibson, S. 2015. *Animal, vegetable, mineral? How eighteenth-century science disrupted the natural order*, Oxford, Oxford University Press.

Glazer, A. M. 2016. *Crystallography: A very short introduction*, Oxford, Oxford University Press.

Kubbinga, H. H. 1990. Hélène Metzger et la théorie corpusculaire des stahliens au XVIIᵉ siècle. *In*: Freudenthal, G. (ed.) *Études sur / Studies on Hélène Metzger*, Leiden, Brill.

La Hire. 1710. Sur une espèce de talc qu'on trouve communément proche de Paris au-dessous des bancs de pierre de plâtre. *Histoire de l'Académie royale des sciences, avec les mémoires de mathématique & de physique*, 341–352.

Lalande, A. 1899. *La dissolution opposée à l'évolution dans les sciences physiques et morales*, Paris.

Lalande, A. 1947. L'involution. Autobiographie philosophique. *Les études philosophiques, nouvelle série*, 1, 1–10.

Lamarck, J. B. P. A. D. M. D. 1914 [1809]. *Zoological philosophy: An exposition with regard to the natural history of animals*, [S.l.], Macmillan.

Lémery, N. 1680 [1675]. *A course of chymistry, containing the easiest manner of performing those operations that are in use in physick. Illustrated with many curious remarks, and useful discourses upon each operation. Together with additional remarks to the former operations, the process of the volatile salt of tartar, and some other useful preparations by way of appendix*, London, Printed for Walter Kettilby at the Bishop's head in St. Paul's church-yard.

Lévy-Bruhl, L. 1910. *Les fonctions mentales dans les sociétés inférieures*, Paris, Alcan.

Lévy-Bruhl, L. 1925 [1922]. *La mentalité primitive*, Paris, Alcan.

Metzger, H. 1918. *La genèse de la science des cristaux*, Paris, Alcan.

Metzger, H. 1922. Letters to Sarton. *Sarton papers bMs Am 1803 (1032)*, Cambridge, MA, Houghton Library.

Metzger, H. 1923. André Cresson: *Les réactions intellectuelles élémentaires*. Isis, 5, 473–474.

Metzger, H. 1926. *Les concepts scientifiques*, Paris, Alcan.

Metzger, H. 1927a. André Cresson, *Les courants de la pensée philosophique française. Isis*, 9, 489–490.

Metzger, H. 1927b. Letters to Sarton. *Sarton papers bMS Am 1803 (1032)*, Cambridge, MA, Houghton Library

Metzger, H. 1927c. Lucien Lévy-Bruhl, *L'âme primitive. Isis*, 9, 482–486.

Metzger, H. 1929. *Les fonctions mentales chez les sociétés inférieures* by Lucien Lévy Bruhl; *Les non-civilisés et nous, différence irréductible ou identité foncière* by Raoul Allier; *La raison primitive* by Olivier Leroy. *Isis*, 12, 343–347.

Metzger, H. 1930. La philosophie de Lévy-Bruhl et l'histoire des sciences. *Archeion*, 12, 15–24.

Metzger, H. 1932. *Le surnaturel et la nature dans la mentalité primitive* by Lucien Lévy-Bruhl; *La mentalité primitive. Isis*, 17, 450–453.

Metzger, H. 1935. Lucien Lévy-Bruhl, *La mythologie primitive. Le monde mythique des australiens et des papous*. Paris 1935. *Archeion*, 17, 104–106.

Metzger, H. 1938. *Attraction universelle et religion naturelle chez quelques commentateurs anglais de Newton*, Paris, Hermann.

Metzger, H. 1969 [1923]. *Les doctrines chimiques en France du début du XVII$^e$ à la fin du XVIII$^e$ siècle*, Paris, Blanchard.

Metzger, H. 1974 [1930]. *Newton, Stahl, Boerhaave et la doctrine chimique*, Paris, Blanchard.

Metzger, H. 1987. *La méthode philosophique en histoire des sciences. Textes 1914–1939, réunis par Gad Freudenthal*, Paris, Fayard.

Metzger, H. 1991 [1930]. *Chemistry*, West Cornwall, CT, Lucust Hill Press.

Metzger, H. 2019 [1936]. The a priori in scientific theory and the history of science (Appendix 3).

Metzger, H. 2019 [1937]. The philosophical method in the history of science (Appendix 4).

Osborn, H. F. 1918. *The origin and evolution of life on the theory of action reaction and interaction of energy*, [S.l.], Bell.

Polanyi, M. 1958. *Personal knowledge: Towards a post-critical philosophy*, London, Routledge & Kegan Paul.

Polanyi, M. 1966. *The tacit dimension*, Garden City, New York, Doubleday.

Spencer, H. 1900 [1879–93]. *The principles of ethics*, Edinburgh, Williams and Norgate.

Spencer, H. 2003 [1897]. *The principles of sociology*, New Brunswick, NJ, London, Transaction Publishers.

Todériciu, D. 1974. Sur la vraie biographie de Samuel Duclos (Du Clos) Cotreau. *Revue d'Histoire des Sciences*, 27, 64–67.

Wallerius, J. G. 1753. *Minéralogie, ou description générale des substances du règne mineral. Ouvrage traduit de l'Allemand par Paul Henri Thiry baron d'Holbach*, Paris.

# 2    The study of science in the making

If we want to grasp the true meaning of the succession of scientific theories, we will have to venture into the troubled, agitated, obscure regions of science in the making

(Metzger, *Les doctrines chimiques en France du début du XVII<sup>e</sup> à la fin du XVIII<sup>e</sup> siècle*)

## Science in the making and thought in its nascent phase

Metzger's *La chimie* (Metzger, 1991 [1930]) is a popular history of chemistry, and a very short book compared with *Les doctrines chimiques* (Metzger, 1969 [1923]), *Newton, Stahl, Boerhaave* (Metzger, 1974 [1930]) and *La genèse de la science des cristaux* (Metzger, 1918). It gives us far fewer insights in Metzger's historiography and theory of mind than her other books. Nevertheless, its structure tells us a great deal about what she thought was important in the history of science. The first part, in which she discussed the formation of chemistry between the Renaissance and the beginning of the nineteenth century, occupies well over two-thirds of the whole book. This part, in her words, covers the period that ends with 'the epoch when the general principles currently taught in our high school classes were imposed by scientists' (Metzger, 1991 [1930], p. 95). In other words, she dedicated most of her history of chemistry to periods in which chemistry, as her readers knew it, did not exist. Indeed, she wrote:

> chemistry was developed mostly during the first half of the nineteenth century, with the discoveries of elements unknown until then, with the creation of compounds or series of compounds, with diffusion of atomic weights of the former, with composition of the latter, and in general with an attempt to classify rationally all of them.
>
> (Metzger, 1991 [1930], p. 125)

She concluded her work with a short chapter on the 'latest magnificent discoveries' (Metzger, 1991 [1930]).

Although she emphasised the importance of nineteenth-century research for current chemistry, in her work as a historian she largely ignored the recent past, and focussed on earlier periods. It was not uncommon for historians of chemistry to start their narrations well before the nineteenth century. In fact, other historians of chemistry, as for instance Marcellin Berthelot (Berthelot, 1889, Berthelot, 1893c, Berthelot, 1893b, Berthelot, 1893a), and Maurice Delacre (Delacre, 1920), covered earlier periods than she did. Delacre's history of chemistry begins with a section on 'primitive age', which precedes one on ancient Egypt (Delacre, 1920, Chapter 1).[1] However, as mentioned in Chapter 1, she aimed to write history of a kind different from Delacre's 'triumphalist' and anachronistic narrative (Metzger, 1974 [1930], p. 7, fn.1); (Halleux, 1990, p. 36). She had resolved not to project current ideas onto the past; her focus on the early-modern period was not rooted in a desire to force early ideas and practices into a long, continuous and homogeneous narrative. She disagreed with the shape of that type of narratives, but also with their contents. Already in her first book, *La genèse de la science des cristaux*, Metzger launched her criticism of a history of science as history of scientific results (Metzger, 1918). Her history of chemistry books that followed remained faithful to her initial approach, and *La chimie* is no exception. As Bensaude-Vincent has put it, '[t]he fact of writing a general popular treatise seems in fact to strengthen her resolve not to write a history of discoveries' (Bensaude-Vincent, 1987, p. 75). Metzger studied periods in which chemistry and crystallography were not fully formed as disciplines, and, as far as crystallography is concerned, periods that were overlooked by her contemporaries. As Christine Blondel points out, the crystallographers of her time started the history of their discipline with Haüy, that is where Metzger concluded *La genèse* (Blondel, 1990, p. 214). The unusual timeframe of her history of crystallography, and its rationale, were also noted at the time. The *Isis* reviewer of *La genèse* commented that she went further back in time than other historians had done, in order to find the first 'sketches' of the 'final work', i. e. crystallography as a properly formed science (Guinet, 1921).Why did she think that a study of activities and theories around matter in the seventeenth and eighteenth centuries would be valuable? What would we learn from it?

Metzger studied the early periods of crystallography and chemistry so that she could better observe 'science in the making' [*la science en formation*]. Science in the making has two related meanings. The first is the formation of disciplines; in her case crystallography and chemistry. She followed the tortuous route by which chemistry and crystallography have finally become disciplines with their own practitioners who regard themselves, and are seen by others, as chemists and crystallographers. Related to this first sense, there is a second sense of science in the making: this is research in the early phase of intuition, first hypotheses and tentative manipulation of substances. This is what Metzger called the 'nascent phase of thought'. Metzger described it as follows:

[T]hought in its nascent phase, thought in the moment that it is formed within us, thought that arises in the thinker at the precise moment when, in a manner of speaking, it awakes within him. I use the expression 'thought in its nascent phase' in the same way that one talks of the nascent state of hydrogen in the laboratory, active hydrogen that frees itself instantly when hydrochloric acid attacks zinc, as opposed to the less active hydrogen stored in an ordinary container.

(Metzger, 2019 [1937])

Thought in its nascent phase is present in scientific research in any time; indeed, she wrote that it was at 'the origin of all progress in scientific theory'(Metzger, 2019 [1937]). It can be detected in modern science, although not through the examination of well-polished publications and textbooks. A history of results for her will never capture science as an activity, indeed as a human activity that cannot be separated from human beings' rich, complex and varied lives and minds. The historian of science for her should rather aim to access live research, and the thoughts and practices of scientists at the beginning of a new project. Although thought in its nascent phase is present in the live experience of modern science, it can be more easily observed in the study of nature before it becomes crystallised into a proper modern discipline. The lack of a homogeneous point of view and of an agreed method left scholars free to follow their intuitions and spontaneous inferences. In those early times, expansive thought pervaded research in a way that would not be possible later on, when much of the scientists' efforts are spent in reflecting on acquired notions and theories, checking their logical coherence, import and consistency with other parts of knowledge. As a historian and as a philosopher Metzger wanted to plunge into the supposed confusion and obscurity that the *Encylopédie* had imputed to chemistry[2]. As a historian, she valued the study of science in the making because she aimed to trace the various paths and detours that have led to chemistry and crystallography. As a philosopher, she valued the study of thought in its nascent phase because for her it is by observing those moments of great creativity that we gain better access to the secrets of our minds, and their rational and non-rational faculties. Indeed, as we have seen in Chapter 1, her study of scientific concepts is similarly focussed on the phase in which concepts emerge and take shape, rather than when they are already well defined and universally accepted. She made that choice because she was convinced that the early formation of concepts, just as of theories and disciplines, affords us a special access to the workings of the human mind.

## The disunity of science in the making

At the time when Metzger was writing her works, Karl Popper declared that it was not 'the business of epistemology' to analyse the 'initial stage' of the scientist's work, namely, 'conceiving or inventing a theory'. For him, that was rather the business of 'empirical psychology' (Popper, 1999 [1935], pp. 7–8). By contrast, Metzger's object, as a historian and as a philosopher, was precisely that initial stage of formation of concepts and theories which Popper declared irrelevant to epistemology. Meanwhile, logical empiricists were promoting the ideal of the unity of science, based on method and language. Just as Metzger organised and attended the conferences of the International Academy for the History of Science, Otto Neurath successfully proposed a series of International Congresses for the Unity of Science, and directed, alongside Philip Frank and Charles Morris, the International Institute for the Unity of Science (Cat, 1998). We shall see later in the present book that Metzger opposed the Vienna Circle's research projects, which were destined to spread widely, and dominate philosophy of science for decades to come. Here, it is relevant to note that she did not think that the unity of science was a reality or an ideal. Moreover, philosophically it was rather the study of variety that was of value. Indeed, she set out to study a myriad of texts, ideas, experiments and practices, which for her showed a dazzling disunity of the research into nature. This disunity is apparent at a number of levels. First, there is social disunity, involving individuals who made a living from their work as well as curious gentlemen who sought beautiful objects for their cabinets. The scholars whom she considered relevant to the formation of crystallography included mineralogists, as well as 'anatomists, medics, and amateurs of all sorts' who were struck by those strange 'stones'; they observed and studied them in different ways and to different ends. Metzger depicted some practitioners as mineralogists who aimed at describing their minerals without recourse to laboratory work, others as chemists who used analysis and synthesis in order to classify the same minerals (Metzger, 1918, p. 79). The former group did not share with the latter either practices or methods. Metzger refused to write a history of isolated 'great men' and chose to work in what she called the 'chaos' of eighteenth-century study of minerals (Metzger, 1918, p. 23). Similarly, she plunged into the 'anarchy' of early seventeenth-century iatrochemistry for her study of the development of chemistry in France (Metzger, 1969 [1923], p. 143). She followed isolated and scattered scholars, who sometimes did not even aspire to be part of a community. For the earlier periods she studied, this social disunity required on the one hand the consultations of many texts, and on the other an awareness that individuals could be guided by significantly different methods, or indeed no method at all. She pointed out that it was only with the establishment of the academies first in Italy, then in England, France and Germany, and later in Scandinavian countries, that scholars ceased to be isolated (Metzger, 1991 [1930], p. 33). She emphasised the importance for science of the type of collaboration that the academies offered. For instance, she argued that in the late seventeenth century, the 'Florence Academy' [Accademia del Cimento] might not have fostered any new discovery, but it was nevertheless crucial for promoting a model of experimental rigour. As a consequence, scholars were less likely to let themselves be 'dazzled' by some unlikely idea; they rather aimed to carry out proper experimental work (Metzger, 1918, p. 132). For her, in the second part of the eighteenth century, science became increasingly more social: even relatively isolated scholars regarded themselves as part of a community, and as able to give a small contribution to a shared effort (Metzger, 1918, p. 118).

The social variety of people engaged in the activities that for Metzger contributed to the formation of chemistry corresponded to a variety of practices. Those practitioners' activities were remarkably diverse, and so were their aims and concerns. Not only pharmacists, medics and alchemists were involved in transformation of matter, but also artisans who made a variety of objects, including glass, porcelain, vinegar and candles (Metzger, 1991 [1930], p. 4). The level of specialised knowledge required for engaging in various activities varied, but also increased with time. The experiments of Paracelsian medics were simple, and anybody could perform them; they were in fact straightforward observations from which the scholar extracted a symbolic meaning, by using generalisations and analogies (Metzger, 1969 [1923], p. 439). However, Metzger pointed out that this simple method was already criticised by van Helmont (1579/1580–1644), before falling out of favour at the time of Lémery, because it could not be reconciled either with Descartes' mechanistic philosophy, or with Gassendi and Boyle's atomism. Nevertheless, Metzger also emphasised that experimental habits do not get replaced at once with theory change; in fact, they often survive in some quarters (Metzger, 1969 [1923], p. 350). Experimental practices were often more diverse that theories, although some of them lasted longer than theories. As mentioned in Chapter 1, when medics and pharmacists added to their vegetable and animal products mineral ones, they continued to employ the same techniques and the same tools as they had done previously (Metzger, 1969 [1923], p. 347). On the other hand, the role of new tools, or of new uses of existing tools, cannot be underestimated. The introduction of the microscope, which directed scholars to pay attention to very small particles that form crystals, had an enormous impact on crystallography (Metzger, 1918, p. 54).

The process of the creation of a science involves the making of its objects. In the nascent phase of a science, its objects are not clearly defined; they may vary from locality to locality, group to group, and even practitioner to practitioner. Chemistry may be the study of matter and its transformations, but what does 'matter' include, in different times and across places? Metzger explained that some scholars regarded as their object of study the entire universe, whereas others only studied metals, and others focussed on the human body. Moreover, for many early scholars, the study of matter was about shedding light on 'the metaphysics of creation'; the Bible, and Genesis in particular, were an important source of natural knowledge (Metzger, 1991 [1930], p. 4). As we have seen in Chapter 1, crystallography only slowly defined its object; indeed, up until the early eighteenth century, scholars neither defined crystals nor distinguished them from 'angular stones' that were not crystals. It was only thanks to Moritz Anton Cappeller (or Kappeler, 1685–1769) that crystals were classified according to their composition and formation. Cappeller distinguished three types of crystals as far as their composition is concerned: stones, metals and salts. He also presented six classes of crystals based on their formation, including evaporation and distillation. Even more importantly for Metzger, he aimed to provide a comprehensive description of crystals based on the general physical mechanism of crystallisation, beyond the variations depending on their compositions and formation (Metzger, 1918, p. 43); (Cappeller, 1723). The variety of practitioners, methods, tools and scientific objects is reflected in theoretical disunity. For Metzger the theoretical fragmentation was first counteracted by Cartesianism, which proposed a strict theoretical discipline (Metzger, 1969 [1923], p. 291). However, for her it would be wrong to think that chemistry had then reached a theoretical unity. In fact, chemists interpreted Descartes' philosophy very freely, and indeed combined it with ancient theories. She quoted Daniel Duncan, author of a treatise on natural chemistry and animal nourishment published in 1681,[3] who was a fervent admirer of Descartes' philosophy, and employed the language of mechanistic philosophy for his own theories. These theories, however, also included the Paracelsian assumption of the analogy of macrocosm and microcosm, among other ancient beliefs (Metzger, 1969 [1923], pp. 241ff.). Even when Descartes' mechanistic philosophy apparently prevailed, chemists adapted it to their specific problems, and in particular to the chemical reactions that they observed. Metzger cited clear examples of the different positions taken by supporters of mechanistic philosophy: François André proposed that there are two principles of the natural world – acids, whose particles are pointy, and alkalis, whose particles are hollow. Bertrand, appealing to the same mechanistic philosophy, rejected the acid-alkali dualism by arguing that it is at odds with nature's simplicity. Similarly, Boyle refused even to consider this dualism (Metzger, 1969 [1923], p. 273).[4] The variations and adaptations of mechanistic theory were so significant that, despite their superficial resemblance, they ended up being incompatible with one another (Metzger, 1969 [1923], p. 274). In later periods, notably after chemistry was transformed by the success of Newton's theory, theoretical disunity lessened, but did not disappear. In fact, for Metzger, Newtonianism offered to the chemist a synthesis between ancient ideas of action at a distance, reminiscent of active analogy, and the corpuscular structure of matter of atomism (Metzger, 1974 [1930], p. 52). In this light, Newtonian chemistry was already a mixture of two approaches, although in a more organised and coherent form than previous bricolage. Although Newtonianism brought back previous ideas against the 'too great' simplicity of the Cartesian and atomistic mechanisms, it avoided the 'excesses of the previous age' (Metzger, 1974 [1930], p. 51). In any case, the success of Newtonianism did not translate in a complete theory change, nor did it offer a comprehensive theory kit to the chemist. Indeed, she found several traces of the old chemistry in Stahl's work, despite its author's open rejection of it. For her, Stahl accepted without discussion assumptions of the previous era, including the porous boundaries between animal, vegetable and mineral realms. Similarly, his view of vital phenomena as independent from mechanical reactions was an important leftover of previous approaches, notably iatrochemistry. She concluded that Stahl's violent attack on iatrochemistry, rather than showing his distance from it, is evidence of his intellectual roots (Metzger, 1974 [1930], pp. 155–116); (Metzger, 1991 [1930], p. 39). These examples also show that theoretical disunity was not only a feature of science in its nascent phase as a whole, when groups with different theoretical approaches worked alongside one another. Individual projects also showed a lack of theoretical coherence and unity. Ancient ideas did not invariably die once and for all; they rather lived on, or even resurfaced in a different context. For example, Paracelsians held that the 'salt principle' was present in all bodies (Metzger, 1974 [1930], p. 157). This belief translated into the enduring view that salt is a constituent part of metals, a notion that survived the demise of Paracelsian philosophy. Johann Kunckel (1630–1703), like other renowned chemists, held this old-fashioned belief, which however did not prevent him from being very successful in his experiments (Metzger, 1991 [1930], p. 37).

Metzger's narrative of chemistry in its nascent phase presents a science marked by profound disunity, which lessens with time, but does not disappear. Certainly, isolated scholars collaborated more and more closely, notably in the academies. Tools and procedures became progressively more standardised: the balance that was optional in the seventeenth-century laboratory became one of the most important tools in all laboratories after Lavoisier (Metzger, 1991 [1930]).[5] Theoretical disunity also lessened with Cartesianism and then Newtonianism. In short, chemistry progressively became a discipline, with its practitioners, places, instruments, procedures and theories. Yet, Metzger argued that, in the periods she studied, no firm unity and homogeneity was achieved. Similarly, the historian could not detect a linear development towards a set of ideas and attitudes. Different and contrasting ideas persisted; innovators still carried, sometimes unconsciously, assumptions from previous and incompatible theories.

## Science in the making and the human mind

Metzger showed us the complexity of science in the making at every level. She never attempted to eliminate contradictions and inconsistencies from her narrative. In fact, she emphasised the mixtures of old and new assumptions and ideas in the same theory, as well as the diverse activities that benefited the development of chemistry. She considered scholars who would not have regarded themselves as having shared goals or belonging to the same group, especially as far as the earlier periods are concerned. Why then consider these disparate activities together? What united them? Metzger appeared to think that these various activities, theories and scholars were diachronically united by their contribution to modern chemistry. This view presents us with some historiographical problems that I shall discuss in the next section. Here, I shall consider the synchronic perspective. Is the historian justified in considering those diverse activities and theories as a sufficiently coherent object of research? Metzger believed that her own way of reading sources enabled her to isolate their common traits. What was her reading method? She argued that there are three ways of reading a text. The first is 'contemplative': the reader only considers it in its finished form, i.e. the form that the author thought fit to be presented to the public. This is arguably the type of reading that Metzger rejected most comprehensively, as it does not aim to go beyond the text or to understand the conditions of possibility of the ideas contained in it. Contemplative reading does not seek to capture thought in its nascent form, but only extract facts that historians can place in their pre-established idea of historical development. In fact, she did not conceal her irony when she explained that this is the reading that enables the historian to place a text on the 'royal road of progress' by measuring its distance from the preceding and future stages of science (Metzger, 1938, p. 12).[6] The second approach to a text is the 'polemical point of view'. The polemics to which Metzger referred are not between the reader and the author of the text. Rather, they are the polemics in which the author of the text engages. The reader is concerned with the explicit defence that authors make of their own theories against those of their predecessors and opponents, and with the way in which they manage to overcome resistance and secure the success of their own views (Metzger, 1938, p. 12). Metzger criticised this second manner of reading past texts as well. For her, this type of reading relies on what authors explicitly write about their own methods, and the way in which they want their works to be read. Like the first, this second method is based on the finished product, and on the authors' well-polished presentations of their own theories. Metzger did not spare this second approach from her irony either. She commented that the historians and philosophers who employ the polemical reading generally base their findings on the analysis of the prefaces and conclusions of the works they study, in other words on those parts in which the author 'peremptorily declares how he has worked and how one must work'. Philosophers and historians, she continued, have inferred the method used by scientists from the latter's formal presentations. Metzger conceded that this type of reading has some merits: it offers a coherent presentation of theories, and explains how the authors of some theories have convinced others to change their ideas and attitudes. However, this second type of approach to past texts, which Metzger saw as somewhat superficial and 'formalistic' (Metzger, 1938, p. 12), could not reach the goal that she assigned to history of science, namely understanding the mind. In order to achieve this goal, she proposed a third reading method, which rather focuses on 'the *active* and fecund thought' in its 'nascent state' (Metzger, 1938, p. 12). 'Active thought' for her creates a whole 'monument', which subsequently must undergo strict mathematical and experimental tests. This second phase, however, was not Metzger's central interest. As we know, she aimed to capture the more spontaneous and creative moments of the mind at work. For her, it is the observation of the mind working with fewer constraints than the reflective phase imposes that places the philosopher in the best position to understand it. Her teacher Léon Brunschvicg thought that the history of mathematics gave the philosopher the best access to the workings of the mind, as for him mathematics is the 'supple and living' instrument (Brunschvicg, 1922, p. 591) that enables human intelligence to 'break free' from the limited sphere of sense data, and grasp the structures of reality (Brunschvicg, 1912, p. 10). For him, non-mathematical ideas about nature belong to superstition and religion rather than science, and they are rooted in pre-scientific realism, this being simply naïve, or 'dogmatic', like Aristotle's realism (Brunschvicg, 1934, p. 135); (Brunschvicg, 1922, p. 610). Metzger pursued a similar goal, namely knowledge of the mind, but she chose very different documents. What interested her was not the mind's emancipation from experience, but rather its independence from the strictures of codified procedures and indeed from mathematics. Her practitioners developed their ideas working in a laboratory or attempting to cure a patient, in the midst of life, as it were. The examination of their ideas for her would shed light on the core characteristics of human minds. Unlike Brunschvicg, she did not think that spontaneous thought was opposed to scientific knowledge, but rather that it was the root from which the latter develops.

The texts that Metzger studied were authored by particular scholars, but she did not aim to examine their individual minds, if not as examples of broader attitudes. As she put it in a letter to George Sarton, she did not want to write a history of great men, but rather capture 'the average opinion' of a time (Freudenthal, 1990, Metzger to Sarton, Letter of 21.8.1923, p. 251). The history of great men of science corresponds for her to the history of finished theories and clearly defined concepts. Just as the latter eliminated all the painstaking and uncertain work that has made it possible for widely accepted concepts and theories to emerge, so the history of great men eliminates all the practitioners without whom no scientist would be 'great' (Metzger, 1969 [1923], p. 8). Her history of science in the making is a history of diverse and diffuse practices, of attempts and false starts as well as brilliant intuitions, of some influential scholars as well as numerous obscure practitioners and amateurs. In *La genèse de la science des cristaux*, Metzger explained that, in order to understand the ideas that circulated at the end of the eighteenth century, the historian should not stop and admire the brilliant ideas of an isolated scholar (Metzger, 1918, p. 23). She followed her own advice in her works, in particular in *La genèse* itself and in *Les doctrines chimiques*. She examined as large a number of texts as she could, including neglected authors, who, as she admitted, did not directly contribute to future science. What she found was that the disunity and diversity of nascent science hid some 'guiding trends' that those scholars followed unconsciously (Metzger, 1918, p. 43). Some of these 'trends' can be identified with the concepts that she systematised in *Les concepts scientifiques* (Metzger, 1926), as we have seen in Chapter 1. As we shall see in Chapter 3, she also included emotions and habits as part of the object of the historian, namely the minds of past scholars. The disunity that science in the making exhibits in its tools, practices, social settings and aims finds a partial unity at the deeper level of organising concepts,[7] metaphysical assumptions and worldviews. The amateur and the professional pharmacist, the isolated scholar and the member of an academy can therefore be studied together by the historian as examples of similar, although not identical, minds, which share 'tendencies' or, as Metzger also called them 'mental *a prioris*'. It is only at that level that for her we can find a unity of sorts, which should not mask the 'constant mobility' (Metzger, 1974 [1930], p. 7) of scientific theory in the early and confused stages of a discipline in formation.

As she put it, she aimed to bring back to life the souls of contemporaries of the texts she studied (Metzger, 1974 [1930], p. 8). Since she aimed to study a time and a milieu, she did not primarily attempt to grasp the mind of the author but rather the minds of his disciples, and of the readers of his work. As mentioned in Chapter 1, she lamented that she could not be one of Lémery's pupils, working with him in his laboratory. Similarly, her declared approach to the study of Lavoisier was to be like a 'disciple of this master' (Metzger, 1935, p. 7). Readers may be more removed from the author than his students, and their backgrounds, abilities and aims may differ quite considerably from his and theirs. What did they understand? How did they use those texts? These were her questions. These readers may well have read the same text differently. In her words:

> If a text supports several interpretations, it is not impossible that these interpretations had actually been given by different commentators on the same scientist. Indeed, it is often the case that the evolution of theories moves forward due to this dissociation of fundamental doctrines.
> (Metzger, 2019 [1937])

Contemporary readers, however, would have shared the *a prioris* that she, as a historian, aimed to reconstruct. These *a prioris* would have not only been shared by natural philosophers and their readers, but also by larger groups in the society. Consequently, Metzger believed that the historian must 'read the greatest numbers of similar texts of the same period', for the following reasons:

> 1) To ensure that he has not made a serious mistake in his interpretation and if need be, rectify it; 2) To absorb better the fundamental attitudes

that determined the tendencies of the mentality of the authors under study; 3) To be able to provide a general framework containing each of these attitudes and tendencies of mentality as a particular case.

(Metzger, 2019 [1937])

Not only did natural philosophers share their worldviews and mental *a prioris* with their contemporaries, but they also had aims which a modern reader might deem external to science. To them, however, these aims were part and parcel of their task. For instance, Metzger wrote:

van Helmont's work ... is a product of early seventeenth century Counter-Reformation. By viciously attacking Aristotle's metaphysics and logic, by attacking astrology, he was acting as a scientist who was ridding chemistry of a number of superstitions, but he was also acting as a Christian fighting against the offensive return of paganism.

(Metzger, 2019 [1933])

What Metzger proposed was a circular method. An extensive reading of texts would enable the historian to detect the assumptions and organising concepts of a whole epoch and society. On the other hand, the understanding of those assumptions and concepts would enable her to interpret texts and theories as particular cases of that epoch and society.[8] Her focus on general, if complex, mental 'tendencies', on concepts, aims and practices that belonged to a milieu rather than to an individual, also led to her lack of interest in biography. Indeed, in *Les doctrines chimiques*, she declared that 'the history of chemical theories, as we have attempted to narrate them, is independent of the history of scientists' lives'. She continued by explaining that she had included neither biographical information in her book nor 'any assessment of the psychology of the authors under study', a practice, she noticed, that other historians of chemistry had attempted to use. Her neglect of biography was negatively noted by George Sarton in his review of *Les doctrines* (Sarton, 1924). In her next book, *Newton, Stahl, Boerhaave*, she restated that her focus was on chemical theories rather than on scientists as individual persons, and for this reason she had not discussed the scientists' lives. Her response to Sarton's criticism was to add short biographies at the end of her volume, and to direct the reader to biographical encyclopaedias for further information.[9] As Bernadette Bensaude-Vincent has put it, even in a popular book like *La chimie*, biographical information is relegated in the Appendix, because Metzger refused to do history of great men. Rather, she provided a sort of 'identikit' of the early seventeenth-century chemist (Bensaude-Vincent, 1987). She was resisting what was a mainstream method employed by historians of science, including those close to her, notably George Sarton. She objected to a history of science that to her appeared to be a construction of a large repository of information without philosophical ambitions. By contrast, her history of science was meant to understand the mind and the nature of knowledge.

## The historian and science in the making

In her books on chemistry and crystallography in the making, Metzger chose a large number of sources, which, in her own presentation, exhibited no unity. The *Isis* reviewer of *La genèse* noted that she included a vast number of sources and philosophies in her history, as she linked the history of crystallography to the history of humanity, and focussed on society rather than individuals (Guinet, 1921). Her selection of sources was dictated by her historiography. For instance, she opposed the mainstream view that chemistry started with Lavoisier. The authority to which she chose to appeal may appear as unusual as some of the minor sources in her books: she reached back to the chemist Eugène Chevreul (1786–1889), not particularly famous even in his own time as the author of a history of chemistry (Chevreul, 1866). She emphasised that Chevreul did not put Lavoisier 'on a pedestal', but rather explained how 'this genius' could make chemical knowledge so quickly: he drew, better than others, the consequences of Newton's philosophy of matter (Metzger, 1987 [1932], p. 161). Metzger here aimed to vindicate the role that past theories and practices have played in the formation of modern scientific knowledge. At the same time, she did not present a linear narrative, as her fragmented picture of science in formation did not allow it.[10] In fact, she harshly criticised simple progressive narratives that for her could only be constructed by making a present-centred choice of sources: the perfectly linear histories of science were the histories of great men and great discoveries. However, her project too appears to depend on a present perspective, as it hinges on the role that those scattered, and profoundly different, early sources played in later science. She wrote about the 'formation' of the science of crystals and of chemistry. On the one hand, she aimed to interpret her sources as a contemporary would have done. On the other, she presented them as the early history of current disciplines, and often emphasised the impact of apparently lapsed theories and practices on later science. These two aims may appear contradictory. Was she committing disciplinary anachronism by interpreting varied and disjointed practices and theories as the early history of chemistry and crystallography? Was her whole construction in contradiction with her historical method? Needless to say, past scholars did not know what happened next, and she avoided 'with the utmost care' to read them in the 'blinding light of our current theories' (Metzger, 1974 [1930], p. 6). However, scholars employed, discussed, criticised and rejected previous theories and practices. As a consequence, for her it is the historian's task to find those connections, and understand what moved a particular author or experimenter to innovate, or indeed go back to older practices, or again to mix different practices. Thus, for her it is equally wrong to see Lavoisier's work as a rootless beginning, and to read previous theories simply in the light of his discoveries. For instance, she claimed that, although Lavoisier relied on previous ideas and practices, he did not rely on Jean Rey's view that historians found so similar to his. A century before Lavoisier, Rey explained that tin and iron increase in weight after calcination because the air mixes with calx (Metzger, 1991 [1930], pp. 28–29). Rey's hypothesis, however, was rejected by his contemporaries, and was not developed; it only attracted attention after Lavoisier's discoveries. The suggestion that there is a development between Rey and Lavoisier, for Metzger, is simply an instance of anachronism. Rey's hypothesis may look similar to Lavoisier's, but the former, like his contemporaries', did not connect combustion with calcination, which they saw as separate phenomena. Like his contemporaries, Rey believed air to be a simple substance; he also followed the Aristotelian theory of the four elements. Moreover, he did not confirm, indeed had no means to confirm, his hypothesis experimentally; he could not show that calcination does not take place in the absence of air. Rey's hypothesis was not even new: Metzger pointed out that Rey himself cited other practitioners who had made the same hypothesis. He left his hypothesis on a list with other hypotheses, without expressing a definitive view (Metzger, 1987 [1939], 84). For Metzger, this example illustrates the errors that historians who do not grasp the mentality of a period almost inevitably make: they are struck by superficial similarities and ignore the proper historical 'filiation of theories' (Metzger, 1987 [1939], p. 87). For her, they use these similarities in order to construct anachronistic narratives, centred on the concept of 'precursor'. Metzger sarcastically described a precursor as a prophet who brings to humanity a truth in a very obscure manner, and announces that a great man will come, who will be able to grasp this truth in all clarity (Metzger, 1987 [1939], 79). Her commitment to historical faithfulness prevented her from seeing anybody as a forerunner; the contemporaries of so-called forerunners could not have regarded them as such. Rhetorically, she asked the audience of her talk on this theme if they could see themselves as precursors of some future discovery (Metzger, 1987 [1939], p. 83).

Some of the reasons for the similarities of intuitions that we can find across time for Metzger lie in the natural tendencies of the human mind. The spontaneous use of certain organising concepts can guide minds to go along paths that look alike. We have seen that, for Metzger, Newton relied on the same intuition as Renaissance medics, namely active analogy, because active analogy is a way in which human beings spontaneously organise empirical data. This does not mean that every theory would use it; in fact, Cartesian philosophers rejected it.[11] The historian, for Metzger, must be able to distinguish between the similarities due to the use of a concept that is always at the disposal of human beings, and historical links between theories. Yet, in her presentation of science in its nascent state, Metzger included many texts that have hardly any link with modern science. These works, by her own admission, belong to lapsed history. Why did she then include them in the history of the formations of chemistry and crystallography? She believed that they had contributed to the formation of these sciences, although one may find no direct links between current science and those past theories. For her, they were intermediate steps, as it were, that formed the tortuous path of the emergence of theories and practices that in turn opened the possibility for others to emerge, in a chain that would eventually make it possible for chemistry and crystallography to become fully fledged disciplines. For her, the 'apparently sterile' works of Cartesian and Newtonian scholars prepared the path to a new physics, by 'furnishing science with a great number of ingenious insights and interesting arguments' (Metzger, 1926, p. 226). She defended the role that the 'mystic' van Helmont played in the development of modern chemistry in several ways. At face value, van Helmont's view that all bodies are made of water does not seem promising for the development of chemistry, nor does his claim that there is a universal dissolvent, the 'alcahest', that can return all bodies to their watery state. Yet, Metzger found significant novelty in his work. In van Helmont's time, chemists had a choice between regarding their mixts as formed by the four elements of ancient astronomy (fire, air, water and earth) as the Peripatetics did, and looking for the Paracelsian three principles of salt, mercury and sulphur. Van Helmont, wrote Metzger, sided with neither, nor did he try to reconcile the two theories, as some of his contemporaries did. His intellectual independence for her opened the way for the 'sceptical' attitude that was going to be Boyle's.[12] In fact, she emphasised that an analysis of Boyle's 'apparently revolutionary' theory showed the structure of van Helmont's spiritualist chemistry, which lay at the bottom of the former's 'uncompromising corpuscularism'. She saw the continuity between the two for instance in the crucial belief that there is no absolute metaphysical difference among the diverse substances that we experience, a view that both Peripatetics and Paracelsians would have not entertained. Metzger also saw a positive development in van Helmont's opinion that air played no role in chemical reaction, and in particular in combustion. Her affirmation might look paradoxical, but she explained that in so doing van Helmont, by denying that air 'feeds' fire, downplayed the role of fire as well. He undermined the previously clear notion of fire, which could no longer be accepted without further analysis. According to Metzger, this was a stepping stone in the subsequent differentiation of the theories of light or heat and of combustion. Other times, scholars whose theories were discarded for her still left a mark in chemists' practices. To stay with van Helmont, she argued that his use of the balance to measure quantity of matter was taken up by Borrichius, and signalled a new approach that would be as we know crucial for chemistry.

Metzger also included in her histories ideas that were not taken up by others either in their times or later on. She did so for various reasons. Notably, from a historiographical point of view, ideas that have been fought against were crucial to her narratives, as she did not write a history of results, but rather a history of how theories were formed. From a philosophical point of view, she analysed discarded theories and practices as examples of ways of thinking and doing that give access to past minds. In other words, for her the task of reconstructing past worldviews and mental *a prioris* require to examine as broad a set of documents as possible: unsuccessful hypotheses and theories could provide insights on what it was possible to conceive. She looked at a deeper level than positive theories in order to find order in the disunity of early science. Despite her great efforts to be true to the worldview of the authors of her sources and their contemporary readers, her general aim in writing about science in the nascent state may still be seen as anachronistic. Indeed, only by positioning herself in the present could she consider selected seventeenth- and eighteenth-century theories and practices as the nascent phases of chemistry and crystallography. In order words, Metzger the historian took as the guiding principle of her narrative something that her sources could not have known, namely that sciences in the modern sense would have emerged out of the patchwork of their theories and practices. As much as she aspired to be a contemporary of her metallurgists and medics, she placed herself at the end of her narrative, and her position gave meaning to the whole construction. She was aware of her partial present-centredness, and indeed defended it. She aimed to read past texts as faithfully as possible, but, on the other hand, as a historian, she created narratives that did not belong in the past. Crucially, she wrote for her own contemporary readers, not unlike a bridge between past and present.[13]

## Notes

1 The long narrative is also present in English-language histories of chemistry of the same period: see J.R. Partington's history, published a few years after Metzger's, which also starts from ancient times (Partington, 1970 [1937]).
2 See Chapter 1.
3 (Duncan, 1681); Metzger had a 1682 Montpellier edition in her bibliography (Metzger, 1969 [1923], p. 473).
4 François André, author of *Entretiens sur l'acide et l'alcali* (André, 1672); he wrote a second edition of his work after Boyle attacked him (Boas, 1958, p. 89). Bertrand, whose first name is not known, was a physician from Marseilles (Clericuzio, 2000, p. 176). Metzger discussed his treatise on acids and alkalis (Bertrand, 1683).
5 See Chapter 1.
6 In his review of Metzger's *Les doctrines chimiques*, Sarton pointed out that Metzger did not follow a 'royal road' in explaining chemical discoveries, but rather a longer route. He continued with his metaphor, by saying that Metzger did not only stop at capital cities, but also in the countryside, and in small villages (Sarton, 1924, p. 58).
7 I take the expression, and the concept of, 'organizing concept' from Ian Hacking (Hacking, 1999).
8 We shall see more in detail how she proposed the historian should understand past minds in Chapter 3.

9 (Metzger, 1974 [1930], pp. 309–322, p. 11). J.R.R. Christie has analysed the rhetorical device that Metzger employed here: her invitation to consult biographical encyclopaedias in fact discourages the reader from doing so, as it is clear that Metzger did not regard biographies as important (Christie, 1987, p. 102).
10 I shall discuss Metzger's type of historical narrative in Chapter 4.
11 The Cartesians officially and consciously rejected analogy; however, as we have seen in Chapter 1, Descartes himself employed it when he advised Huygens to use glass rather than diamonds in his experiments.
12 I refer, of course, to Boyle's *Sceptical Chymist* (Boyle, 1964 [1661]).
13 Regarding the partial present-centredness of Metzger's historiography, see Moro-Abadía (2008), Moro-Abadía (2009).

## References

André, F. 1672. *Entretiens sur l'acide et l'alcali*, Paris.

Bensaude-Vincent, B. 1987. Hélène Metzger's *La chimie*: A popular treatise. *History of Science*, 25, 71–84.

Berthelot, M. 1889. *Introduction a l'étude de la chimie des anciens et du Moyen âge*, Paris, Steiheil.

Berthelot, M. 1893a. *La chimie au Moyen âge, tome I: essai sur la transmission de la science antique au Moyen âge*, Paris, Imprimerie nationale.

Berthelot, M. 1893b. *La chimie au Moyen âge, tome II: l'alchimie syriaque*, Paris, Imprimerie nationale.

Berthelot, M. 1893c. *La chimie au Moyen âge, tome III: la chimie arabe*, Paris, Imprimerie nationale.

Bertrand. 1683. *Réflexions nouvelles sur l'acide et sur l'alcali: où après avoir démontré que ces deux sels ne peuvent pas être les principes des mixtes, on fait voir le véritable usage qu'on en peut faire dans la physique & dans la médecine*, Lyon, Amaurly.

Blondel, C. 1990. Hélène Metzger et la cristallographie: de la pratique d'une science à son histoire. *In*: Freudenthal, G. (ed.) *Études sur / Studies on Hélène Metzger*, Leiden, Brill.

Boas, M. 1958. *Robert Boyle and seventeenth-century chemistry*, Cambridge, Cambridge University Press.

Boyle, R. 1964 [1661]. *The sceptical chymist*, [S.l.], Dutton.

Brunschvicg, L. 1912. *Les étapes de la philosophie mathématique*, Paris, Alcan.

Brunschvicg, L. 1922. *L'expérience humaine et la causalité physique*, Paris, Alcan.

Brunschvicg, L. 1934. *Les âges de l'intelligence*, Paris, Alcan.

Cappeller, M. A. 1723. *Prodromus crystallographiae de crystallis improprie sic dictis commentarium*, Lucerne Typis Henrici Rennwardi Wyssing.

Cat, J. 1998. Unity of science. *Routledge encyclopedia of philosophy*, Taylor & Francis. Retrieved 14 May 2019 from https://www.rep.routledge.com/articles/thematic/unity-of-science/v-1.

Chevreul, E. 1866. *Histoire des connaissances chimiques*, Paris, Guérin et Morgand.

Christie, J. R. R. 1987. Narrative and rhetoric in Hélène Metzger's historiography of eighteenth century chemistry. *History of Science*, 25, 99–109.

Clericuzio, A. 2000. *Elements, principles and corpuscles: A study of atomism and chemistry in the seventeenth century*, Dordrecht, London, Kluwer Academic.

Delacre, M. 1920. *Histoire de la chimie*, Paris, Gauthier-Villars et cie.

Duncan, D. 1681. *La chymie naturelle, ou l'explication chymique et méchanique de la nourriture de l'animal*, Paris, Jean d'Houry

Freudenthal, G. (ed.) 1990. *Études sur / Studies on Hélène Metzger*, Leiden, Brill.

Guinet, L. 1921. Hélène Metzger, *La genèse de la science des cristaux. Isis*, 3, 445–446.

Hacking, I. 1999. Historical meta-epistemology. *In*: Carl, W. & Daston, L. (eds.) *Wahrheit und Geschicte*, Göttingen, Vandenhoeck & Ruprecht.

Halleux, R. 1990. Visages de Van Helmont, depuis Hélène Metzger jusqu'à Walter Pagel. *In*: Freudenthal, G. (ed.) *Études sur / Studies on Hélène Metzger*, Leiden, Brill.

Metzger, H. 1918. *La genèse de la science des cristaux*, Paris, Alcan.

Metzger, H. 1926. *Les concepts scientifiques*, Paris, Alcan.

Metzger, H. 1935. *La philosophie de la matière chez Lavoisier*, Paris, Hermann.

Metzger, H. 1938. *Attraction universelle et religion naturelle chez quelques commentateurs anglais de Newton*, Paris, Hermann.

Metzger, H. 1969 [1923]. *Les doctrines chimiques en France du début du XVIIᵉ à la fin du XVIIIᵉ siècle*, Paris, Blanchard.

Metzger, H. 1974 [1930]. *Newton, Stahl, Boerhaave et la doctrine chimique*, Paris, Blanchard.

Metzger, H. 1987 [1932]. Eugène Chevreul historien de la chimie. *In*: Metzger, H. 1987. *La méthode philosophique en histoire des sciences. Textes 1914–1939, réunis par Gad Freudenthal*, Paris, Fayard.

Metzger, H. 1987 [1939]. Le rôle des précurseurs dans l'évolution de la science. *In*: Metzger, H. 1987. *La méthode philosophique en histoire des sciences. Textes 1914–1939, réunis par Gad Freudenthal*, Paris, Fayard.

Metzger, H. 1991 [1930]. *Chemistry*, West Cornwall, CT, Lucust Hill Press.

Metzger, H. 2019 [1933]. Should historians of science become the contemporaries of the scholars they study? (Appendix 1).

Metzger, H. 2019 [1937]. The philosophical method in the history of science (Appendix 4).

Moro-Abadía, O. 2008. Beyond the whig history interpretation of history: Lessons on 'presentism' from Hélène Metzger. *Studies in History and Philosophy of Science*, 39, 194–201.

Moro-Abadía, O. 2009. Thinking about 'presentism' from a historian's perspective: Herbert Butterfield and Hélène Metzger. *History of Science*, 47, 57–77.

Partington, J. R. 1970 [1937]. *A history of chemistry, Vol.1*, London, New York, Macmillan, St. Martins Pr.

Popper, K. 1999 [1935]. *The logic of scientific discovery*, London, Routledge.

Sarton, G. 1924. Hélène Metzger, *Les doctrines chimiques en France du début du XVIIe à la fin du XVIIIe siècle. Isis*, 6, 57–64.

# 3 Emotions, habits and sympathy

## The mind, nature and history

Metzger challenged the view that alchemical and early chemical texts were just records of irrational musings. She argued that if we grasp the concepts that guided past scholars' reasoning, their inferences will often appear perfectly, and sometimes imperfectly, good. In *Les concepts scientifiques* (Metzger, 1926), she systematically analysed the concepts that she had extracted from her sources. For her, only familiarity with past scholars' assumptions and concepts would enable the historian to understand early-modern natural philosophies. In fact, for her, any type of knowledge, including modern science, can only make sense if one is acquainted with the relevant assumptions and concepts. In her philosophical training, she absorbed Brunschvicg's neo-Kantianism, which, as I have argued elsewhere, was crucial to the formation of the philosophical movement that we have come to call historical epistemology (Chimisso, 2008). This is the context in which she represented knowledge of the natural world as a combination of *a priori* forms and experience. In her words:

> human intelligence produces light at the same time as being an organ of vision, just like the phosphorescent eyes of deep-sea fish; if one prevents it from providing its own brightness, it quickly becomes blind.
>
> (Metzger, 2019 [1936])

As mentioned in the previous chapters, Metzger called all the *a priori* forms simply '*a prioris*'. She claimed that her own meaning of '*a priori*' was broader than the definition that one could find in Lalande's Dictionary of Philosophy (Lalande, 1999 [1926]). What she did not mention is that instead of using *a priori* either as an adjective or as an adverb, she employed it as a noun. With this proviso in mind, here is how she presented it:

> First, and in virtue of my working method, I shall be obliged to expand on the meaning of '*a priori*' slightly beyond the definition provided by

Mr Lalande's admirable *Vocabulary*, which I shall quote to you first of all. One calls *a priori* notions that are independent of experience, 'at least relatively speaking, that is to say that experience presupposes them, and does not suffice to explain them, even when they only apply to experience'. Mr Lalande further adds: '*A priori* does not indicate chronological precedence but rather logical precedence'.

Now, since we do not wish to meddle with this excellent definition, we shall instead have to adapt it to the way of seeing of the historian of science who, over the course of his work, has acquired the conviction that certain characteristics of theories derive as much from experience and observation as they do from the researcher's mentality. If one acknowledges this, the *a priori* will not just represent all the concepts in place prior to experience and on which experience relies. The *a priori* will also represent the fundamental tendencies that produce these concepts.

(Metzger, 2019 [1936])

It is important to note that she did not consider only concepts, not to mention the fully formed and reflectively employed concepts of mature sciences. Rather, she aimed to investigate what she called the 'fundamental tendencies' of the human mind. These 'tendencies' include the mind's spontaneous activity of grouping objects together into classes, which is behind scientific concepts (Metzger, 1926, Introduction). The knowledge of those concepts is necessary to understand past sources, but it is not sufficient. Mental *a prioris* also include emotions, habits, passions and desires.[1] She did not think that science in general, and science in the making in particular, could be understood by only appealing to scientific rationality. Whereas in Chapter 1 I have focussed on her study of concepts, here I shall examine the roles that she assigned to emotions, passions and habits in the creation of scientific knowledge. I shall also discuss her view that, like scientific knowledge, historical knowledge cannot solely rely on reason. Notably she defended, quite originally in her milieu, the role of sympathy in historical knowledge, and in particular in the understanding of past texts. Her inclusion of sympathy among the historian's tools has rightly prompted comparisons between her historical method and that of the hermeneutical tradition. I shall however emphasise the differences, because these show her complex view of the relationship between historical and scientific knowledge. As I shall discuss in the last section of this chapter, the roles that she assigned to emotions, passions and habits in scientific knowledge and to sympathy in historical knowledge resulted in a very interesting view of the relationship between these two types of knowledge. She took scientific knowledge to be the model of historical knowledge, but at the same time she was highly critical of positivist historiography, this being in the style of Hippolyte Taine or Henri Berr.

## Emotions, passions and habits

Metzger did not provide a catalogue of emotions and habits in the same way as she did of concepts. In fact, *Les concepts scientifiques* is her only attempt at any systematisation. Although I use the terms 'emotions' and 'passions', I do not even suggest that she had any theory about what these are. Rather, I aim to explain that she assigned an important role in science to non-rational faculties. She also acknowledged the importance of motivations and ambitions that may well be rational, but that are not internal to scientific rationality. Her account of the varied roles of emotions, passions and also habits in the development of knowledge about nature is best found in her narratives and analyses of sources. She appealed to the actors' passions and emotions to explain all stages of investigation into nature. Although she admitted that there is no way to know how human beings first related to nature and represented it, she still imagined them as confused by the extreme variety and extraordinary complexity of natural phenomena (Metzger, 1926, p. 3). For her, the human mind instinctively attempts to order natural objects and phenomena into classes (Metzger, 1926, p. 2), in order to make sense of them. Metzger supposed that the study of nature proceeds from a psychological need, namely the 'shock' resulting from the resistance that nature presents to our understanding. However, she did not indulge in imagining the first steps of humanity, but rather focused in what she saw as the first steps of chemistry and crystallography.

As discussed in Chapter 1, Metzger argued that the alchemists' quest to transform base metals into gold was not irrational if we understand their assumption that nature has a tendency to develop towards perfection, and if we follow their inferences based on analogy (Metzger, 1969 [1923]). She did not deny that pride, ambition, and the desire to be part of an admired sect must have played a role in the alchemists' persistence in an enterprise that consistently met with failure. However, Metzger was not particularly interested in those 'external' passions that simply attracted scholars to a certain discipline or professional group. She rather focused on those passions that suggested concepts to researchers and guided them in their study of nature. And she found passions guiding research alongside rational inferences, or, more precisely, mixed with them. Alchemists and medics may have correctly drawn conclusions from their premises. However, their desire to explain everything in terms of mutual influences, and to rely completely on this premise, also played an important role. For instance, Metzger emphasised that most seventeenth-century medics, in their desire to discover all possible analogies, simply generalised from those they thought they had established. She remarked that the study of science in formation provides ample evidence of many scholars' irresistible desire to discover immediately the homogeneity of the world (Metzger, 1969 [1923], p. 162). In the same period, anatomists, medics and amateurs were guided by their imagination, and for this reason they searched for natural objects that seemed exceptional and attractive. Due to their drive to satisfy their imagination, some of them sought, collected and studied 'stones' that had remarkable shapes, sometimes similar to parts of living beings, or that were charmingly transparent and regular. The aesthetic appeal of those objects was an important element of the beginning of the study of crystals, however unsystematic. Feelings played a fundamental role in the explanation of phenomena, notably in Renaissance philosophy; they were not deemed to belong to the researcher, but rather to objects themselves. For those scholars, gold and aqua regia 'love' each other, and this is why they unite, and 'hold' each other; substances stand in relations of friendship or mutual revulsion, and this is why they combine or fail to do so. The projection of emotions onto minerals was not simply discarded when Renaissance philosophy fell out of favour. In fact, Metzger pointed out that Boerhaave employed the same language, though he did so metaphorically, rather than literally (Metzger, 1974 [1930], pp. 51–52). More generally, for Metzger the concept of 'mutual attraction' is a 'psychological notion' that Becher, Stahl and the latter's disciples borrowed from Renaissance vitalist mysticism: for them, 'atoms of the same species look for each other and are responsible for a good deal of the phenomena studied in chemistry' (Metzger, 1991 [1930], p. 41). Cartesian philosophy led chemists and medics to ignore phenomena that could not be explained in a mechanical way, and to propose a corpuscular view of the structures, activities and transformations of the different parts of an organism. Metzger explained that, according to Cartesian philosophy, the knowledge of the various shapes of the molecules should have been sufficient to explain all phenomena in a living being. She quoted Sprengel (1766–1833),[2] who welcomed this view only in so far as it would eliminate the 'habit' of introducing hidden qualities. These, for him, did not offer valid explanations (Metzger, 1969 [1923], p. 243). Metzger, however, did not fail to point out the non-rational aspects of the enthusiastic reception of Cartesianism. She claimed that Cartesian chemists employed their imagination in order to represent the shapes of the various corpuscles (Metzger, 1969 [1923], p. 243). She also continued in her quotation of Sprengel who lamented the negative consequences of Descartes' 'passion' for mathematics, which led his followers to attempt to explain everything in mathematical terms. Indeed, the reasons why scholars accept some theories rather than others for Metzger may involve motives that are not internal to the theory, and indeed proceed from emotions including fear of the crushing complexity of nature. She explained that in the mid-seventeenth century, many people were irresistibly seduced by mechanistic philosophy because it reduced to unity the great diversity of phenomena, and because it appeared to be consistent with the experimental empiricism that the most famous scientists supported (Metzger, 1969 [1923], p. 251). The desire of simplification of the 'shocking' variety of natural phenomena for Metzger has often driven the explanation of natural phenomena. Newton's physics was for Metzger eminently successful in the simplification and unification of phenomena, as it showed the universe as a harmonic whole. This in her view was an important reason why so many scholars readily accepted Newtonianism: for her they felt a strong 'emotion' in having the harmony of the universe revealed to them (Metzger, 1938, p. 205).

Metzger proposed that in early science expansive thought prevails, and reflective thought plays a smaller role than it does in mature science. Reflective thought for her reconsiders the ideas that expansive thought has produced and submits them to logic and experimental checks. This may seem to suggest that emotions, desires and the imagination should play little or no role in mature science, as the latter is dominated by reflective thought. However, Metzger's view of early and mature science is far less clear-cut than that. Unlike some of her contemporaries, including Léon Brunschvicg, she did not see the development of science as a progressive rationalisation, nor did she interpret as setbacks returns to a larger use of imaginative notions. Newtonianism is arguably the best illustration that Metzger offered of the persistence of expansive thought within a science in which reflective thought plays a crucial role. We have seen in Chapter 2 that Metzger presented Newtonianism as a mixture of Renaissance philosophy and Cartesianism. The former in her presentation was the triumph of expansive thought, offering a wealth of ideas and theories largely based on instinct, emotions and desires, but also on rational inferences from its assumptions. Cartesian philosophy, by contrast, aspired to strict rationalism, relying as it did on mathematics, and excluding anything that was not mechanical, such as the active analogies that had pervaded Renaissance thought. For Metzger, Newtonianism was a synthesis of the two. She argued that the foundations of Newton's system were to be found at two very different levels of mentality: the high level of the reflective thought of Cartesian rationalism, and 'the deep level' of expansive thought (Metzger, 1938, p. 10). Universal attraction could not be accommodated within Cartesian mechanicism. Rather, for Metzger it appeared to have its roots in theories 'as old as the world': its 'psychological origin' lay in the 'mysterious *participations*', which Lévy-Bruhl held to be frequent in primitive thought (Metzger, 1938, p. 9). In other words, Newton for her had 'received and mathematicised' the notion of 'active force of desire' that is characteristic of instinctive thought, and had informed Renaissance philosophy (Metzger, 1938, p. 140). For her, rationality and emotions, strict logic and free imagination, are components of the study of nature at all times, though in different proportions and in different ways.

Similarly, she regarded habits as an important part of science. As she often reminded her readers, chemistry is above all doing: experiments involve procedures that may be learned independently of any particular theory. It may be easy to overlook the importance that she attached to experimental habits if we limit ourselves to her theoretical essays. In 'The philosophical method', she goes as far as saying that:

[W]hen I speak of the history of science, I am talking about the history of scientific thought and nothing else; the other aspects of science, which include observation, experimentation, measuring, calculation procedures or techniques for building laboratory equipment, either have no importance, or have only secondary importance, or are an aid to thought or the creation of thought ... in my opinion and to cite a specific example, all the reagents that one encounters in the labelled bottles on the shelves of chemistry laboratories, all the instruments that one finds in these same laboratories are embodied products of a theory which are used in order to be able to verify this theory, but which must be understood and in any case can only be understood in relation to this theory.

(Metzger, 2019 [1937])[3]

The above statement seems to downplay the importance of experimentation and practices, and to deny them any autonomy from theory. However, I think it is necessary to read it charitably, and in context. Her talk was polemical, as her audience, and in particular Henri Berr, the director of the Centre de synthèse, favoured a positivistic view of historical research, focused on facts and dates. By contrast, for Metzger the historian should aim to understand past minds. Reagents and instruments have to be understood within the context of the mentality of those using them. In that sense, they are 'theory-laden', as we would say nowadays. However, in her other writings, Metzger did not claim what appears to be suggested by the above quotation: first, that experimental tools and practices play a small role in the history of science, and second that scientific objects and instruments should be interpreted in terms of a particular scientific theory. In fact, as I have discussed, she attached great importance to the instruments that scholars employed, in order to understand what they could or could not do. Metzger expressed the wish to visit past laboratories in order to work alongside past chemists and be their disciple. This was of course an unreachable aim, but she kept it as an ideal to which she tried to get as close as possible when studying the documents at her disposal. As far as the autonomy of observations and experimental practices from theory is concerned, we should make a distinction. She certainly thought that objects and instruments should not be studied separately from the minds of the scholars who used them. What they observed, as well as their tools, for her were shaped by their view of the universe, although, as I discussed in Chapter 1, she also showed that technological limitations – for instance the impossibility of obtaining pure substances and keeping them at a constant temperature – had an impact on their objects. At the same time, on several occasions she presented some experimental practices as independent of a particular theory. Moreover, she suggested that the enduring use of some experimental methods was not connected with rational choices, but rather with habit. A good example is the way in which she discussed the use of the balance in the chemical laboratory.

She noticed that chemists did not wait for the theoretical justification for its use, which came when Newton defined quantity of matter as mass. In fact van Helmont, Borrichius, Boyle, Lémery and Jean Rey, to mention but a few, already used the balance. However, what is particularly interesting is that the Cartesians did so as well, although according to their theory quantity of matter was supposed to coincide with extension rather than weight (Metzger, 1974 [1930], pp. 23ff). In other words, the experimental use of the balance pre-dated the theory that justified it, and remained in place even at a time when Cartesian theory contradicted it.

The force of habit appears also to be behind what Metzger described as the unreflective use of past methods and ideas within new theoretical and philosophical contexts. A case in point is the manner in which certain beliefs of Paracelsianism survived theory and method change. As mentioned, Metzger explained that Paracelsian scholars used simple observations from which they extracted a symbolic meaning that in turn enabled them to make generalisations. By employing analogy, they could infer the system of the world from simple observation. This method, however, fell out of favour already in Lémery's time; moreover, Francis Bacon was critical of the chemists' generalisations. Then Descartes' mechanicism and Boyle's atomism excluded analogy, which was at odds with their philosophies. These theoretical and methodological changes may suggest that no trace of Paracelsian ideas should have survived in later practice. Yet, Metzger claimed that traces of Paracelsus' ideas could be found in documents for a long time; her more recent example was from 1746, when La Garaye (1675–1755), a student of Lémery (Lehman, 2014), could preface his *Chimie hydraulique* with assertions that belonged to a previous era. Namely, he put forward a general view of the composition of all mixts that was in fact suggested by the distillation of plants. For Metzger, those generalisations belonged to a previous age, and they were repeated by pharmacists, who did not really believe in them, by force of habit (Metzger, 1969 [1923], pp. 350–351).

### The historian's task: milieu, time and active sympathy

Metzger attached great importance to the study of the general culture of a period in order to understand her chemical texts, as we shall see in Chapter 4. This should not come as a surprise, as we know that for her the *a prioris* of the scholars she studied belonged to their culture as a whole. But does a certain milieu inevitably produce a certain type of science or natural philosophy? Metzger was acquainted with the work of Hippolyte Taine, the historian who saw time and milieu as the creators, as it were, of minds. Taine proposed a strong causal link between social conditions and type of art. For him, the form of a society, the conditions of servitude or liberty, wealth or poverty, and its religion, would engender 'corresponding' desires and sentiments (Taine, 1865). Indeed, he famously claimed that 'race', 'milieu' and 'epoch' determine the moral status of a people (Taine, 1877 [1864], p. 10); (Taine, 1865, pp. 158–162). When Metzger commented on his view, she omitted 'race', as for her 'innate and hereditary dispositions' (Taine, 1877 [1864], p. 10) belonged to the human race as a whole, and arguably she preferred to ignore Taine's remarks on the subject. Milieu and epoch, however, were important for her understanding of mentality. Her first encounter with Taine's *Philosophy of Art* had left her indecisive (Metzger, 1987 [1930]). She agreed that milieu and time played an important role in the development of art and, she added, of science. However, she raised doubts about Taine's determinism, which led him to claim that there exist laws regulating the type of culture that could appear in a certain moment in a given place (Taine, 1877 [1864], p. 17). Metzger in 1930 refused to give a definite verdict on the matter, all the same suggesting that the evolution of science, literature and arts could be 'partially' determined by the 'moment'. Seven years later, she was ready to express her judgement more explicitly. On the one hand, she acknowledged the importance of milieu and epoch for 'thought in its nascent phase', that is the object of her study. She conceded that thought depends to a great extent on the circumstance in which it emerges. On the other hand, she rejected Taine's notion that thought could be a direct and inevitable product of its milieu. Indeed, she held that thought in its nascent phase was one of the factors that contribute to cultural and social changes (Metzger, 2019 [1937]). In other words, for Metzger the 'thought' that she aimed to describe is both 'created' and 'creator' (Metzger, 2019 [1937]). If thought were solely created, indeed causally determined by 'milieu and time', a positive knowledge of the latter would afford the knowledge of their effects; in her case, science in its nascent phase. However, she held that the new ideas that human minds create are able to influence the course of events. This made her task more complex, as past thought should then be understood both as a product of its time, and as a producer of change. Consequently, she held that the study of milieu and time is important, but not sufficient in order to understand past science. She needed to capture the living thought processes of past scholars, which cannot be reduced to instances of their culture. How could this be done? How is the historian to understand not only the conceptual structure of past scholars' thought processes and their metaphysics, but also their emotions, their unexpressed desires, their unreflexively acquired habits? How can the historian capture the passions that drove past scholars, the ambitions that motivated them and the imagination that inspired them?

Metgzer's works show that she took great care in textual analysis, as well as in the identification both of tools and reagents, and of scientific concepts. She was also attentive to the social condition of scholars, in particular to whether they were isolated amateurs or celebrated members of academies. For her, scholarship is 'an indispensable tool' that 'must be constructed as carefully as if we had the hope that it would provide us with real knowledge all by itself' (Metzger, 2019 [1937]); (Metzger, 1974 [1930], p. 12). For her the most careful scholarship is necessary to understand past texts, but is it sufficient? She thought that scholarship does not provide us with full knowledge of the past. A real understanding of early science can only be achieved by capturing the 'active thought' of past scholars in the moment in which it was formed within them. Undoubtedly, her aim appeared to go beyond the evidence that her texts could provide. She therefore proposed a method to be used in addition to traditional scholarship. In order to present it, she used an analogy with what she knew best: chemistry. For her, the historian should reconstruct past scholars' thought in her own mind, as our mind is 'the only reagent available to us' (Metzger, 2019 [1937]). The scientific image might be surprising, because Metzger proposed that historians should employ a method that is not normally associated with science, although it is in fact consistent with her view of science. This method is 'active sympathy', which for her historians should use in order to re-create past thought in their own minds. In her words:

> In the course of his research, [the historian] will constantly strive to go deeper so to understand the past better, to grasp with more assurance and more active sympathy the creative thought of the past, into which he will breathe new life and which he will, for a brief moment, bring back to life.
>
> (Metzger, 2019 [1933])

She knew that what she proposed was no easy task, but she thought that this method was necessary in order to achieve her aims. If she had chosen to do a 'history of results', she could have relied on scholarship alone. However, her object was science in the making. Therefore, she aimed to grasp thought in its development, indeed the 'soul' behind scientific research. Her objective was to capture the moment in which the diverse scholars of matter whom she studied came up with ideas and started developing them. She wanted to perceive the fascination and wonder that the collectors of 'angular stones' felt in the presence of their beautiful objects. She aspired to reading iatrochemists' treatises as if she belonged to that profession; she imagined being one of Lémery's students, trying her hand at experiments as if not only did she have the same mental categories as them, but also the same aims, emotions, habits and imagination. Her own goal was to understand those scholars in their own terms, as they were carrying out their research. With this aim in mind, she felt that active sympathy was the only option, however arduous and uncertain. She wrote:

> [The historian] knows that his intelligence, the sympathy he desires to have, and his historical imagination are very limited … But he also knows that he does not have any reagents available other than himself; and so, because he realises that for us humans, thought, far from being something that floats around, is always the thought of a thinker; because he knows that he does not want to engage in the vain and pointless game of studying thought as if it were a thing, an isolated and inert object made of words, he rather places himself with timid boldness at the very centre of his effort, like a spider in the centre of its web.
>
> (Metzger, 2019 [1936])

Consequently, she encouraged historians to use all their resources without fear:

> Do not be afraid to make use of all the resources that your creative imagination and inventive intelligence graciously offer you in order to seek out a thought that is impenetrable to our scholarly methods of analysis and synthesis ... do not be afraid to employ all your active sympathy to seek out a theory of which the text gives you the outward appearance and not the soul ...
>
> (Metzger, 2019 [1936])

Metzger may seem to propose a method aimed at capturing thoughts, emotions and experiences of the authors of the texts she studied. However, her goal was not primarily to reproduce the thoughts and possibly the feelings of Lémery, Stahl or Boerhaave, but rather those of a hypothetical contemporary student or follower of them. Although she thought that it was important to establish authors' aims and ambitions, her most important objective was not to understand the thought of an individual, even if this was the author she was studying. Consequently, her 'sympathy' was not directed at a particular person, but rather at a type of person, living in a certain time and place, possibly with a certain position in society and in their profession or trade. This is how Metzger described the historian's 'effort for sympathy':

> When the historian studies the work of Cartesian, Newtonian, Stahlian, and Condillacian chemists, who worked and made their science advance in the eighteenth century, he must make himself in turn into a Cartesian, a Newtonian, a Stahlian, and a Condillacian. He thereby understands, or at least attempts to understand, the different mentalities that have suggested our distant predecessors' theories, hypotheses and experimental research. He manages to understand fully, or at least takes the only path that enables him to understand fully, the work of the diverse schools that accepted their research, or of those that opposed it.
>
> (Metzger, 2019 [1933])

She aspired to turn herself into a generic Newtonian and Stahlian, rather than into a specific Newtonian or Stahlian scholar, let alone into Newton or Stahl.[4] If the historian aimed to find out the intentions and thoughts of the individual author of a text, then that text would generally admit one interpretation, in accordance with the author's view. The reader's perspective that Metzger chose, by contrast, admits more than one interpretation, as we have seen in Chapter 2. Metzger did not think that an epoch had a monolithic mentality; certainly, for her the *a prioris* that regulated past scholars' works were shared by their contemporaries, but there were variations due to a range of facts, including the fragmented social and intellectual circumstances of science in its nascent phase, as seen in Chapter 2. Although the interpretation of a contemporary reader for her would have been within the bounds of certain *a prioris*, it would have not been fully determined by them. This is why different readers could have different interpretations of the same contemporary text. These different interpretations would create new ideas, and possibly changes in the mentalities of their contemporaries, something that Taine would have denied.

## Scientific and historical knowledge

Metzger's introduction of 'active sympathy' as a historical method has prompted comparison with the hermeneutical tradition. Martin Carrier, and especially Gad Freudenthal, has brought out the similarities between her approach and that tradition, so I shall not go over them here (Carrier, 1990); (Freudenthal, 1990). Rather, I shall point out some differences, as they are relevant to the understanding of her method. An interesting point is the relation between history and science. Although in the nineteenth century hermeneutical understanding could be seen as useful in science, for instance by August Boeckh (1785–1867), the human sciences, including history, were widely regarded as opposed to the natural sciences, notably by Johann Gustav Droysen (1808–1884) and Wilhelm Dilthey (1833–1911) (Stueber, 2006, pp. 10–11); (Dilthey, 2002 [1910]). For them, hermeneutics belonged to history, rather than science. By contrast, positivist historians, including Taine, and, in Metzger's milieu, Henri Berr and other historians of the Centre de synthèse, denied the distinction between science and history, as they saw the latter as one of the positive sciences. Like positivist historians, Metzger took the natural sciences as her model and inspiration for the historical method. However, she turned the relationship between science and history on its head. For her, it is not the case that history should exclusively rely on rationality and hard evidence. Rather, she recognised that the natural sciences also rely on emotions, passions and habits. Her study of science in the making was aimed to show that the analysis of fully justified and polished theories does not show the methods and practices employed by scientists when they are innovative and creative. Moreover, just as history for her cannot be history of facts, so science is not an account, or an explanation, of mind-independent facts. She aimed to show that science is a human activity, shaped by what she called mental *a prioris*. Scientific concepts, worldviews, emotions, passions and habits affect and direct scientific research. Scientists' mentalities change, and so does science. In her words:

> the *a priori* is not and cannot always be similar to itself everywhere; or rather, there is not only one *a priori*, but several different *a prioris* that are sometimes heterogeneous and incompatible.
>
> (Metzger, 2019 [1936])

For her, the content of science is historical and does not exist independently of the human beings who make it, and of their diverse ways of thinking.

The objects of the natural science on the one hand and history on the other, however, appear to be different. Georges Canguilhem, in order to explain this difference, chose Metzger's *La genèse de la science des cristaux* (Metzger, 1918) as an example. He explained that the object of the crystallographer is 'given', and has no history. By contrast, the object of Metzger as a historian is 'a discourse on the discourses held on the nature of crystals'. The discourses that are the objects of the historian are in contrast with the object of science because, unlike the latter, they have a history (Canguilhem, 1994 [1968], 16). Although it would be pointless to deny the difference between a crystal and a treatise on crystallography, I think Canguilhem went too far in his distinction, which incidentally is not entirely consistent with his own historical epistemology.[5] This, however, is not my concern here. My topic here is Metzger's work, and Canguilhem's distinction is not reflected in her view of the scientific object. Are crystals completely 'given' and without history? As shown in Chapter 1, Metzger took a lot of care to show that for a long time 'crystals' as we currently understand them were not the objects of the different scholars who wrote about 'angular stones', or simply about rare and beautiful 'stones', Metzger showed her reader the historicity of the objects of natural philosophers and scientists. Scientific objects for her are neither given, nor mind-independent. In order to understand past scholars' objects, she needed to understand their minds. R. G. Collingwood distinguished between the 'outside' and 'inside' of an event, the latter being the actors' thoughts, and argued that 'in … penetrating to the inside of events and detecting the thought which they express, the historian is doing something which the scientist need not and cannot do' (Collingwood, 1994, 214). By contrast, Metzger was very much aware that students of science need to learn how nature is organised according to their teachers, which things are scientific objects, and how these are classified. Indeed, even without deliberate learning, scientists would adopt the view of nature of their own time. In Collingwood's terminology, they would have to learn, or simply absorb, the 'inside', that is the thoughts that had shaped their objects and set their aims. Whether an object is an angular stone or a crystal, whether this same object is in the same category as living beings or separate from them, and whether its shape is meaningful in terms of classification, all depends on human beings' thinking. Metzger aimed to become a 'student' of past scholars so that she would learn those scholars' 'thoughts', just as their own students would have done in order to train as pharmacists, medics, chemists and crystallographers. Despite this and other significant differences between Collingwood and Metzger, the former's proposal of 're-enactment' as a historical method can be usefully compared with Metzger's method. Collingwood proposed that 'the history of thought, and therefore all history, is the re-enactment of past thought in the historian's own mind' (Collingwood, 1994, p. 215). She too aimed to re-enact past scholars' thoughts. However, she also aspired to repeat their experiments, and understand their procedures, as she did not think that a clear line could be drawn between doing and thinking. Her re-enactment (a term she did not use) was cognitive, practical and also sympathetic.[6] For her, science was the product of our cognition, emotions and experimental practices, and therefore the historian had to activate and re-enact thoughts, emotion and experiments. To be fair, I have no evidence that Metzger actually re-enacted experiments, but her attention to how they were carried out, and to the diverse uses of tools and reagents, suggests at least an ideal re-enactment of practical experiences, in addition to cognitive processes.

Metzger appears to have modelled her view of historical practice on scientific practice, which is not surprising, if we bear in mind that her formal education was scientific.[7] However, her view of scientific practice was very different from those she encountered in her milieu or read about. First of all, both science and history for her are not exclusively rational enterprises. Emotions and imagination play a crucial role in both. For her, both scientists and historians formulate hypotheses and follow practices that are informed by their *a prioris* and their worldviews. This is not something that can be eliminated: knowledge cannot be 'purified' of mental *a prioris* but, for her, human beings are capable to reflect on their own assumptions and on their own emotions, and those of past scholars. She presented her views to her fellow scholars at the Centre de synthèse with the following words:

> And certainly, you [historians] will end up making mistakes in interpreting texts ... But who will stop you from checking your suppositions and testing your hypotheses? Who will prevent you from criticising your work yourself by applying the conclusions of your reflections on other texts? If you have got it right, other texts from the same era – it does not matter whether by the same scholar or by other scholars – will suddenly become clear and transparent ...
>
> (Metzger, 2019 [1936])

Moreover, both the scientists and the historians revise their hypotheses as their research expands, and more information comes to light thanks to contact with their peers.

> [The historian] offers his conclusions humbly, he does not seek to impose them; he asks other historians to check or correct his own assertions.
>
> (Metzger, 2019 [1936])

For Metzger, historians, just as scientists, need a community in order to advance their knowledge. Historians correct one another's mistakes and amend individual interpretations. She emphasised the importance of the creation of academies for the advancement of science, as they offered a formal setting for discussion. As a historian and philosopher, she sought communities, at the Centre de synthèse, at the Institut d'histoire des sciences, at the École practique des hautes études, and at the International Academy of History of Science. The large number of book reviews that she wrote are also part of her dialogue with other scholars. The possibility of error for her should not stop historians either from putting forward their theories and views, or from using all the tools at their disposal, including emotions and imagination. This is because their interpretations of the past will always be subject to their own reflective thought as well as that of their peers.

## Notes

1 Iris Van der Tuin has interpreted Metzger's *a priori* by focussing on the creative and empathic side of it; indeed she calls it the 'creative *a priori*' (Van der Tuin, 2013).
2 Kurt Polykarp Sprengel, author of a history of medicine, which Metzger quoted in French translation; she cited a 1924 edition of the following publication, and incorrectly indicated Bosquillon as the translator rather than the editor (Sprengler, 1815–1820). See Broman (1996).
3 Judith Schlager has judged Metzger's focus on scientific thought as fundamental to subsequent historiography; indeed she regards Metzger's work as the 'point of origin' of intellectual history of science (Schlanger, 1990).
4 See on this (Bensaude-Vincent, 1987) and Carrier (Carrier, 1990). Carrier points out that Metzger's emphasis was on 'reception of scientific theories, and that her 'target was never the individual thinker' (Carrier, 1990, p. 139) (p. 139), although in his article he stresses 'author's intentions' (Carrier, 1990, p. 136).
5 For historical epistemologists like Canguilhem, the scientific object is in fact constructed, and as such is historical, as constructions change in history. For an illustration of this, see his history of the concept of reflex (Canguilhem, 1977 [1955]).
6 For the distinction between cognitive and sympathetic, see Jardine (2019), which draws on Wilhelm Dilthey and R.G. Collingwood.
7 Carrier has discussed Metzger's parallel between scientific and historical knowledge (Carrier, 1990).

## References

Bensaude-Vincent, B. 1987. Hélène Metzger's *La chimie*: A popular treatise. *History of Science*, 25, 71–84.

Broman, T. H. 1996. *The transformation of German academic medicine, 1750–1820*, Cambridge, Cambridge University Press.

Canguilhem, G. 1977 [1955]. *La formation du concept de réflexe aux XVIIᵉ et XVIIIᵉ siècles*, Paris, Presses Universitaires de France.

Canguilhem, G. 1994 [1968]. *Études d'histoire et de philosophie des sciences concernant les vivants et la vie*, Paris, Vrin.

Carrier, M. 1990. Some aspects of Hélène Metzger's philosophy of science. *In*: Freudenthal, G. (ed.) *Études sur / Studies on Hélène Metzger*, Leiden, Brill.

Chimisso, C. 2008. *Writing the history of the mind: Philosophy and science in France, 1900 to 1960s*, Aldershot, Ashgate.

Collingwood, R. G. 1994. *The idea of history: With lectures 1926–1928*, Oxford, Oxford University Press.

Dilthey, W. 2002 [1910]. *The formation of the historical world in the human sciences*, Princeton and Oxford, Princeton University Press.

Freudenthal, G. 1990. Épistémologie des sciences de la nature et herméneutique de l'histoire des sciences selon Hélène Metzger. *In*: Freudenthal, G. (ed.) *Études sur / Studies on Hélène Metzger*, Leiden, Brill.

Jardine, N. 2019. Emotional engagement in scientific biographies. *In*: Matravers, D. & Waldow, A. (eds.) *Philosophical perspectives on empathy: Theoretical approaches and emerging challenges*, Abingdon, Routledge.

Lalande, A. 1999 [1926]. *Vocabulaire technique et critique de la philosophie*, Paris, Presses Universitaires de France.

Lehman, C. 2014. Pierre-Joseph Macquer: Chemistry in the French Enlightenment. *Osiris*, 29, 245–61.

Metzger, H. 1918. *La genèse de la science des cristaux*, Paris, Alcan.

Metzger, H. 1926. *Les concepts scientifiques*, Paris, Alcan.

Metzger, H. 1938. *Attraction universelle et religion naturelle chez quelques commentateurs anglais de Newton*, Paris, Hermann.

Metzger, H. 1969 [1923]. *Les doctrines chimiques en France du début du XVIIᵉ à la fin du XVIIIᵉ siècle*, Paris, Blanchard.

Metzger, H. 1974 [1930]. *Newton, Stahl, Boerhaave et la doctrine chimique*, Paris, Blanchard.

Metzger, H. 1987 [1930]. La philosophie d'Hippolyte Taine et l'histoire sociale des sciences. *In*: Metzger, H. 1987. *La méthode philosophique en histoire des sciences. Textes 1914–1939, réunis par Gad Freudenthal*, Paris, Fayard.

Metzger, H. 1991 [1930]. *Chemistry*, West Cornwall, CT, Lucust Hill Press.

Metzger, H. 2019 [1933]. Should historians of science become the contemporaries of the scholars they study? (Appendix 1).

Metzger, H. 2019 [1936]. The *a priori* in scientific theory and the history of science (Appendix 3).

Metzger, H. 2019 [1937]. The philosophical method in the history of science (Appendix 4).

Schlanger, J. 1990. L'histoire de la pensée scientifique et les autres histoires intellectuels. *In*: Freudenthal, G. (ed.) *Études sur/ Studies on Hélène Metzger*, Leiden, Brill.

Sprengler, K. 1815–1820. *Histoire de la médecine depuis son origine jusqu'au XIXᵉ siècle; traduit de l'allemand sur la seconde édition par A.J.L. Jourdan*, Montpellier, Gabon.

Stueber, K. R. 2006. *Rediscovering empathy: Agency, folk psychology, and the human sciences*, Cambridge, MA, London, MIT.

Taine, H. 1865. *The philosophy of art*, London, Baillière

Taine, H. 1877 [1864]. *History of English literature*, London, Chatto & Windus.

Van Der Tuin, I. 2013. Non-reductive continental naturalism in the contemporary humanities: Working with Hélène Metzger's philosophical reflections. *History of the Human Sciences*, 26, 88–105.

# Part II

# The role of other disciplines in Metzger's historiography

# 4 Historical synthesis and the shape of history of science

We cannot believe that chemistry developed independently of general history
(Metzger, 'La littérature chimique française aux XVII$^e$ et XVIII$^e$ siécles').

[T]he history of the science of crystals, like the history of all science, is linked in its broad outlines to the general history of humanity
(Metzger, *La genèse de la science des cristaux*)

## Sources and narrative

For any historian, arguably the two most important decisions concern the selection of sources and the manner in which those should be linked into a narrative. Metzger paid great attention to both problems, although in rather different manners. She often discussed the first problem, which can be further split into two issues. The first issue regards which authors should be considered as part of the history of chemistry and history of crystallography. The second issue is whether social, cultural, economic and institutional information should be seen as relevant, or indeed crucial, to the history of science. I have mentioned Metzger's view concerning the importance of considering texts by so-called minor authors. She fully followed her own advice, as her histories explore sources that other historians would have considered of little consequence. Indeed, in her books she included scholars who were rather obscure in their own times, and scholars who did not find many followers in the subsequent history of science. As mentioned in the previous chapters, her aim to understand the mental *a prioris* of a period and milieu required her to be as comprehensive as possible in her readings. She needed to reconstruct mental categories, worldviews, habits, emotions, places of work and practices, as she thought that the scholars whom she studied shared their mentalities with their contemporaries. Her views seem to suggest that for her the historian should understand the society and culture in which science developed, and therefore use sources external to the disciplines under study. Indeed, she emphasised the importance not only of studying texts outside history of science narrowly defined, but also of becoming familiar with the cultural and social settings in which her sources were produced. However, her approach to the latter issue differs from what a current reader may expect. She did not write social or cultural histories of science, at least not in the current sense. In fact, her current readers may be puzzled by the thinness of the information about the society to which past scholars belonged. Even without expecting from her a type of history of which she had no concrete example in her immediate milieu, one could be struck by the discrepancy between her theoretical claims and her practice. I shall discuss this apparent discrepancy, and its reasons. I shall also contrast her position with that of the senior members of the Centre de synthèse, notably the founder and director Henri Berr, and Abel Rey. The latter, professor of history and philosophy of science at the Sorbonne, founded the Institut d'histoire des sciences where Metzger lectured. Rey was closely involved with the Centre: he directed its Unit for general synthesis, and co-edited its journal *Revue de synthèse*. As we shall see, there was a significant difference between the role that Berr and Rey, on the one hand, and Metzger, on the other, assigned to 'historical synthesis'.

Historians are faced by decisions not only about what sources to include in their work, but also about how to construct a narrative out of diverse documents. As we know, Metzger thoroughly criticised, indeed mocked, the narrative that anachronistically interpreted the past study of nature only as imperfectly anticipating modern science, and as inevitably leading to the latter. However, in her milieu there were several models of progressive narratives that were more sophisticated than that anachronistic history of science. Some of these did not display a linear or continuous development, but nevertheless portrayed the history of science as reason's progressive elimination of superstition, mysticism and error. Notably, Léon Brunschvicg's histories were constructed along those lines. His narratives were certainly progressive, but at the same time exhibited discontinuities, indeed profound breaks. At the time, the issue of whether the history of science is continuous or discontinuous was one of the most contentious historiographical questions, which had its roots in philosophy, and affected most of the emergent disciplines in France. Brunschvicg and his student Gaston Bachelard inaugurated a strong tradition of discontinuous history of science. Metzger also had close contact with Alexandre Koyré, the historian of the scientific revolution. Lucien Lévy-Bruhl proposed a discontinuity between modern and 'primitive' people's mentalities. Within the Institute of ethnology his discontinuist view was at odds with that of the Durkheimians, first of all Marcel Mauss, who favoured a continuous development of the human mind, and considered contemporary non-literate cultures as early stages of a common trajectory that had reached its peak with European modernity. As far as history of science is concerned, the most famous continuist view was due to Pierre Duhem. Metzger confronted that model, just as she did Meyerson's, which was also opposed to the discontinuities proposed by other philosophers and historians. She was very much immersed in those controversies. Yet we do not find a clear pronouncement from her in favour of either view. In her historical work, she was at pains to show the differences that separate the modern reader from past mentalities and metaphysics, and one period under study from another. Indeed, she explained new theories often in terms of changes in dominant concepts and worldviews. On the other hand, she established continuities where other historians did not see them. She had a view of the history of science that escaped the perhaps too easy contraposition of continuity and discontinuity, all the while avoiding a linear narrative. Her dominant aim to be faithful to the past made it difficult for her to interpret past research into nature as a chain, this being either broken or unbroken, of events and theories. She offered a more complex picture, which, with an image borrowed from her presentation of commentaries on Newton, we could call 'theme and variations'. The issue of selection of sources and that of narrative are obviously interlinked as two sides of the shape of history. It would be pointless to wonder whether Metzger arrived at her choice of the type of narrative that she wanted to construct from her sources, or whether she looked for neglected sources because she had already in mind a certain narrative. As a historian, she tended to show that her sources compelled her, but then as a historiographer she easily changed perspective and regarded the selection of sources as a choice that presupposes a particular view of science, knowledge and history.

## Types of sources and their uses

Metzger's first book, *La genèse de la science des cristaux*, is based on primary sources concerning minerals, stones, shells, fossils, mountains, the sea, the whole of nature, and also crystals. A modern reader may well and easily label it as internal history. However, in the Conclusion, she wondered about the social and cultural circumstances that made it possible for particular theories, and indeed disciplines, to develop and succeed. Those circumstances appeared to her to be particularly important for crystallography in its nascent phase, when there were few constraints on methods and objectives. In her words, crystallography in the eighteenth century did not have its own 'personality' and therefore could not develop according to 'an internal necessity' (Metzger, 1918, p. 223). In fact, she did not seem to believe that science ever proceeds by 'internal necessity' alone. We have seen in Chapter 3 that, although she rejected Taine's determinism, still she assigned an important role to social and cultural circumstances in the development of the history of science. In fact, years before discussing Taine, she was already convinced that science is 'partially determined' by the 'human circumstances [*conditions*] of its own development' (Metzger, 1918, p. 224). A modern reader could still wonder why she did not tackle her own questions in her subsequent books, and focus more decisively on external factors. Here there are different questions. First, we should determine what factor is internal or external to science. We shall see in Chapter 7, in relation to Kuhn's view of Metzger's and Koyré's histories of science as internalist, and Kuhn's own definition of external history, that the distinction is not clear-cut, or indeed clear. For Metzger, as for other historians of mentalities, the study of the 'intellectual milieu' of a period, as opposed to a narrow focus on exceptional works (Metzger, 1969 [1923], pp. 11, 12), was considered a departure from a traditional history of science, constructed as a succession of scientific discoveries and theories. Her reconstruction of the metaphysical and religious assumptions behind theories of matter was a conscious effort to widen the focus of her study.[1] As she put it, in order to explain the evolution of chemistry, it would be 'insufficient' to examine chemistry alone. For her one must not neglect the 'powerful' and 'dominant' 'external influences'. Among others, these 'external influences' include the philosophy of the time, and the development of the sciences on which chemistry was modelled, namely astronomy, mechanics and physics (Metzger, 1969 [1923], p. 345).

However, these 'external influences' for Metzger also included social circumstances, such as the creation of academies and all the 'social transformations' that facilitated the dissemination of science, and that brought science to be considered in its practical utility. All those factors, when joined with laboratory work, shaped chemical theories (Metzger, 1969 [1923], p. 345). Indeed, she taught her Sorbonne students that:

> no point of human history is irrelevant for the history of the sciences; social and political history, history of philosophy, history of literature including history of theatre ... history of industry and trade, to which could be also added history of art and history of occultism, are [all] useful to the researcher who proposes to study the chemists' writings.
>
> (Metzger, 1987 [1935], p. 138)

In her works, one can detect that social and cultural considerations were important to her. For instance, she paid attention to cultural practices popular in the higher classes, notably the 'fashion', as she called it, of filling cabinets of natural history (Metzger, 1918, p. 221). It did not escape her that early collectors were wealthy people, who studied for their own pleasure, and did not know 'rigorous and professional discipline'. Their science was 'a topic of conversation' rather than 'technical research' (Metzger, 1918, p. 223). The social success of theories was also something that concerned her. She interrogated herself on the reasons that led some theories to be accepted and some others to be ignored. Once again, we found her view on this dilemma already formed at the time of her first book. Scientific theories for her cannot succeed only on the strength of 'internal' virtues, such as explanatory power or simplicity. Indeed, she hardly mentioned those, and when she did, as we have seen in Chapter 3 in relation to simplicity, she linked them to aspirations and feelings external to the logic of a theory. Rather, she concluded that only theories that are consistent with the general social development achieve success, whereas 'the opinions and methods that are opposed to or on the fringe of mainstream thought, are permanently or temporarily condemned to be forgotten' (Metzger, 1918, p. 224). An illustration of her view is her claim that the immediate reception of Newton's theories should be understood in the context of the rapid 'progress of literature, the arts, the sciences, commerce and industry in England'. This progress would have predisposed Newton's commentators to be optimistic about 'life and humanity' and indeed human ability to understand nature (Metzger, 1938, pp. 196–197). Despite these indications and suggestions, Metzger fell short of providing specific answers to questions about the success or failure of particular theories and practices. She discussed the psychological, metaphysical and religious basis of theories and of their success, but she provided little information about the societies in which those theories emerged, beside general, and indeed infrequent, remarks such as those mentioned above about England. In fact, her type of history of science appeared to be largely based on textual analysis. If anything, her later books are even more focussed on textual analysis than *La genèse*. It appears that on the one hand, she considered the 'external influences' as far as these included philosophy, religion and other sciences, but refrained from analysing the 'external influences' of 'social and political history', despite encouraging her students to do so. Several reasons can be put forward for this apparent discrepancy between her theoretical aspirations and her practice, in addition to the most general one, namely that in her milieu she found no examples of a social history of science. She was well aware that history of science was in its nascent state, just as the sciences that were the focus of her studies. The most urgent task seemed to be unearthing sources, understanding them as well as possible, and constructing a narrative, however provisional. Indeed, just as she raised social and cultural questions about past scholars and theories, she retreated by humbly claiming that her 'fragmentary' work on the genesis of crystallography was not in a position to offer answers (Metzger, 1918, p. 223). She timidly tucked away in a footnote her curiosity about the 'social modifications' that brought about Condorcet's idea of indefinite progress, as opposed to the idea of perfection as a fixed goal that most pre-Enlightenment authors held. She added that an answer to such question would require a more in-depth work than her present study of the emergency of crystallography, and 'a great number of scientific efforts of all kinds' (Metzger, 1918, p. 225). In her more mature work, *Newton, Stahl, Boerhaave* (Metzger, 1974 [1930]), she still emphasised that the history of science was well behind history of art and history of literature. Bearing in mind the infancy of the history of pre-Lavoisierian chemistry, she wrote that her aim in writing that book was to provide a 'basis for discussion' for future historians of science.

Metzger thought that questions about the social and political circumstances of the development of science were important, but also that she was not in a position to answer them. Her involvement with the Centre de synthèse and the Institut d'histoire des sciences should have provided what she needed. The Centre in particular was a forum in which historians and philosophers gave talks and discussed their work. Moreover, both these institutions had historical 'synthesis' as their aim (Berr, 1899, Berr, 1911, Berr, 1914, Berr, 1921, Berr, 1950, Berr, 1953); (Castelli Gattinara, 1997); (Febvre, 1952); (Gemelli, 1987). However, there were profound historiographical differences between Metzger on the one hand, and Berr and Rey on the other, to the point that she even struggled to communicate her view of history to them. Historical synthesis as envisaged by Berr and Rey, with all their differences, was based on a positivistic view of history. Berr's life work was to make history into a positive science, based on the inductive method. In fact, his overall aim appears to be a reduction of all knowledge to the methods and aims of the natural sciences. His historical method, namely historical synthesis, for him would turn history into 'science, true science and full science' (Berr, 1911, p. 23). Berr regarded historical synthesis as a scientific enterprise based on empirical facts, aimed at discovering specific laws. He opposed his synthesis to the philosophical syntheses offered by Émile Boutroux, Émile Bréhier, Léon Brunschvicg and Victor Delbos, and German-style philosophy of history. His view was reflected in the organisation of the Centre de synthèse, which was a hub of production of vocabularies, bibliographies and specialised research carried out in its various units, including that for the history of science. This wealth of information would then be gathered by scholars like Rey, as head of the Unit for general synthesis, and used in order to create a general history (Anonymous [Henri Berr], 1900); (Berr, 1910). Rey's view of history of science was analogous to Berr's view of general history. The proclaimed 'absolute positivism' of his early career (Rey, 1909a, Rey, 1909b) was the basis for his historiographical views. He aimed to make history of science a scientific discipline, based on the empirical method.[2] For him, historians of particular sciences must first gather facts and organise them. These facts would then be coordinated and interpreted, so as to generate a general history of science (Rey, 1930). His programme of a general history of science was realised both in his published work and in the comprehensive syllabi of the Institut d'histoire des sciences, which offered courses on all particular sciences and periods (Anonymous, 1936); (Braunstein, 2006); (Chimisso, 2008 pp. 93ff). The historian of chemistry Aldo Mieli, who headed the Unit for the history of science at the Centre de synthèse, largely shared Berr and Rey's encyclopaedic and positivistic view of history of science, and their focus on bibliographies and recording of historical facts. His pressing preoccupation, at that stage of the development of the history of the sciences, was to fill the gaps in the information available, including bibliographies and biographies (Mieli and Brunet, 1935, p. 18); (Mieli, 1916); (Mieli, 1919); (Bucciantini, 1987). The Centre's accumulation of information and bibliographies had little to do with Metzger's view that 'no point of human history is irrelevant to the history of science', quoted above. Indeed, possibly as a response to the Centre's historians, in the Introduction of *Newton, Stahl, Boerhaave*, she wrote that she did not see the point of 'unearthing all documents and texts' relating to the science of the time. Her aim was rather to 'reconstruct chemical theory' (Metzger, 1974 [1930], p. 12). Metzger had not changed her mind; she had not started defending a narrow view of history of chemistry. Rather, she counterpoised her type of history, focused on mentalities, to the empiricist history of Berr and Mieli, based on accumulation of facts. In fact, Metzger regarded biographies and attribution of discoveries,[3] as well as accumulation of information about the past, when taken to be the historian of science's main task, with contempt.[4] Her ultimate aim was not to establish who the first to come up with a certain idea was, but to capture the creative process of science in the making. She addressed the following words to her fellow historians at the Centre de synthèse:

> To my opponents, I shall only say that if it had been demonstrated that the history of science is only able to satisfy a curiosity that is of course legitimate but philosophically sterile, [a curiosity] that delights in scholarship, picturesque events, elegant patterns of descriptions of theories, [and] from which the creative thought has effectively disappeared ... well then I would immediately cease to devote myself to the history of science.
>
> (Metzger, 2019 [1937])

Moreover, Berr's historical synthesis is similar to what Michel Foucault called 'total history': 'a total description [that] draws all phenomena around a single centre – a principle, a meaning, a world-view, an overall shape' (Foucault, 1972 [1969], p. 10). Metzger, on the contrary, sought to describe the fragmentation of science, as we have seen in Chapter 2, and pointed out the different roles that various mental *a prioris* played in the formation of knowledge.

## Continuous and discontinuous narratives

We have seen that Metzger rejected the perfectly linear and progressive narratives. Her presentation of science in formation is complex and fragmented, as it describes diverse theories and practices side by side. Moreover, a large part of her work is aimed at explaining the 'otherness' of past texts by showing that they were the expression of organising concepts that differed from current ways of thinking, of aims often alien to modern science, and even of habits and emotions that the modern person can only grasp with difficulty. She argued that the study of nature, including modern science, may revive discarded notions and intuitions, but in fact she also paid great attention to lapsed theories and practices that were not picked up again. In short, one could presume that she regarded the history of science as non-linear, discontinuous, and possibly as exhibiting no progress or even no clear narrative. Yet, Metzger presented her books as histories of the 'formation' of the sciences of crystallography and chemistry. She wove diverse sources together in a narrative that implied an extension to reach current science. She showed the provisional end point only in *La chimie*, but her intent was always to narrate the tortuous path that had led to modern science. In other words, Metzger appears on the one hand to emphasise the great differences between early-modern natural philosophy and modern science, and on the other to regard the former as the early stages of the latter. These two views might look incompatible, especially so in her own philosophical milieu. Philosophers and historians of science in her time, and indeed earlier and arguably up to our days,[5] divided on the question of whether history of science is continuous and discontinuous. As François Braunstein put it, the most important difference between Comte and positivist philosophers including Pierre Lafitte and Abel Rey on the one hand, and French historians of science in the Bachelardian tradition on the other, is that the former regarded history of science as continuous, whereas the latter envisaged a discontinuous history (Braunstein, 2008, p. 30). We can certainly extend Braunstein's judgement to Bachelard's teacher Brunschvicg, among others. Moreover, we must not forget the importance of the continuist model proposed by Pierre Duhem, although he never belonged to the Parisian intellectual milieus examined in the present book.

Metzger's acquaintance with these two models was somewhat different from that of those scholars who had a more conventional *curriculum studiorum* than she did. When she wrote *La genèse*, she had not formally studied history or philosophy, although she had close relationships with historians and philosophers. She was married to an academic historian, and her uncle Lucien Lévy-Bruhl, Sorbonne professor of history of philosophy, advised her and read her work. Lévy-Bruhl's opposition of primitive and modern mentalities, when applied to history, as in fact Brunschvicg and Bachelard in revised forms did, suggests a discontinuous narrative. Very soon after writing *La genèse*, she was also exposed to Pierre Duhem's strongly continuist view of history of science. As we know, she submitted *La genèse* as a doctoral dissertation. At her *viva voce* examination, her examiners expressed their surprise that she had not cited Pierre Duhem. She confessed that she had not heard of him, a fact that certainly tells us about her patchy education in history and philosophy (Metzger, 1937, p. 135). She soon filled this gap in her knowledge, and studied Duhem's works. At the same time, however, she also expanded her philosophical knowledge under the guide of Gaston Milhaud and André Lalande (Freudenthal, 1990). Crucially, she attended Brunschvicg's lectures. Brunschvicg, who always regarded himself as a philosopher who 'studied the mind', pursued his philosophical goals largely through the history of science, and mathematics in particular (Brunschvicg, 1923, letter of 2.2.1923); (Brunschvicg, 1922b, Letter to Sarton of 29.12.1922). His histories were focused on the transformation of the mind, pointed to differences in ways of thinking, but were nevertheless constructed as progress. His view of a progressive history, however, was far from the traditional 'triumphal epics' that Metzger mocked. He thought that there was evidence that the mind progressed towards an increasingly more rational state, and a less superstitious and more secular way of regarding nature. However, this progress was neither linear nor continuous. He saw setbacks in philosophies that looked down on mathematics, such as Aristotles', and emphasised the distance between modern and 'primitive' ways of conceiving of nature (Brunschvicg, 1912); (Brunschvicg, 1922a); (Brunschvicg, 1927); (Brunschvicg, 1934); (Brunschvicg, 1947). The progress of rationality for him did not have an end point, as shown by the dramatic transformation of physical theory due to the theory of relativity. For Brunschvicg, the latter was a revolution in ways of thinking, indeed of organising sense data, as it radically revised the concept of time. Brunschvicg opened the path for his student Gaston Bachelard, who elaborated the immensely influential concept of 'epistemological break' [*rupture épistémologique*] (Bachelard, 1951, Bachelard, 1972 [1953], Bachelard, 1986 [1949]). Even before putting forward his famous concept, Bachelard defended discontinuity epistemologically, metaphysically and ontologically in two early books that were attacks on Bergson's concept of *durée* ( Bachelard, 2013 [1931]); (Bachelard, 2000 [1936]); (Bergson, 1998 [1907]). Later on, Metzger also encountered another philosopher who held a strong continuist view of the history of science: Émile Meyerson. Whereas she read Duhem's work only after his death in 1916, and therefore never met him, she had significant personal contact with Meyerson. Like Lévy-Bruhl, Meyerson commented on her drafts, and believed that he guided her in her research. Meyerson's continuist view was one of the main philosophical reasons of his disagreement with Bachelard, who often criticised him (see for instance: (Bachelard, 2002 [1938], p. 19) (Bachelard, 1986 [1949], p. 9)).

Despite their great differences, there was something that Brunschvicg, Bachelard and Meyerson shared in Metzger's eyes: they were all philosophers and, as we shall see in Chapter 5, she thought that they took history less seriously than they should have done. On the other hand, she felt intellectually close to Alexandre Koyré, whom however she met relatively late in her career (Metzger in (Section d'histoire des sciences, 1936)).[6] Koyré's fusion of philosophy and serious history of science was something to which Metzger certainly aspired in her own work. Notably, his lectures that Metzger attended, before standing in for him in his teaching,[7] were on Galileo, as were the papers that he was publishing at the time, and the book he was writing (Koyré, 1935, Koyré, 1936, Koyré, 1937, Koyré, 1939); (Koyré, 1978 [1939]). Both he and Metzger aimed to provide an explanation for the fact that a phenomenon or an idea that looked perfectly reasonable in one period seemed absurd in another period, or was simply not in use. His classical example is the principle of inertial motion, obvious after Galileo, and obviously false for Greek and medieval scholars (Koyré, 1968 [1943], p. 19). In his words:

> why and how [did] the principle of inertial motion, which to us appears so simple, so clear, so plausible and even self-evident, [acquire] this

status of self-evidence and *a priori* truth whereas for the Greeks as well as for the thinkers of the Middle Ages the idea that a body once put in motion will continue to move forever appeared as obviously and evidently false, and even absurd[?]

(Koyré, 1968 [1943], p. 19)

Alexandre Koyré, as he himself recalled, wanted to make sense of change in 'the very framework and patterns of our thinking' that took place with the scientific revolution (Koyré, 1958 [1957], p. v).[8] In other words, Koyré compared the conceptual frameworks of the Middle Ages and of modern people. His discontinuous narrative was based on the comparison of sources from different times. In fact, comparisons based on textual analysis appear to have been the most important methodological tool for continuists and discontinuists alike: all scholars whom I have mentioned here largely based their pronouncements about the continuity or discontinuity of history of science on comparison. For instance, Bachelard, as we shall see more in detail in Chapter 5, compared the 'pre-scientific' mind and the 'scientific mind' that for him were behind early-modern natural philosophy and modern science respectively, and judged that they were radically different (Bachelard, 2002 [1938]). On that basis, and following Brunschvicg's method, he pronounced history of science to be discontinuous.

Comparison of texts was also at the core of continuist views of history of science. A clear example is Duhem, who compared physical theories in different times. According to Aristotelian physics, each element has a natural place where it remains when at rest; only violent action would remove the element from its natural place, and the former will maintain the tendency to return to the latter. This qualitative approach to the study of movement seems completely incompatible with modern physics. Yet, Pierre Duhem argued that, behind this difference, there is a crucial similarity between the two. He wrote that in Aristotelian physics 'a state can be conceived in which the order of the universe would be perfect ... this state would be a state of equilibrium for the world, and what is more, a state of stable equilibrium; removed from this state, the world would tend to return to it ...' (Duhem, 1991 [1914], pp. 309–310). This for him is strikingly similar to the teaching of thermodynamics, according to which all motions and all phenomena produced within an isolated system would make the entropy of the system increase and therefore lead this system to its state of equilibrium. He argued that if one carried out more similar 'comparisons', one would reach the following conclusion:

If we rid the physics of Aristotle and of Scholasticism of the outworn and demoded scientific clothing covering it, and if we bring out in its vigorous and harmonious nakedness the living flesh of this cosmology, we would be struck by its resemblance to our modern physical theory; we recognize in these two doctrines two pictures of the same ontological order, distinct because they are each taken from a different point of view, but in no way discordant.

(Duhem, 1991 [1914], p. 310)

Pierre Duhem compared Aristotelian and modern physics, and judged that a great difference could be disregarded, as more fundamental similarities could be detected; as a result, he established a continuity between the two.

Is comparison a method that can deliver a definitive verdict of the shape of history of science? The historian and political scientist Benedict Anderson spent much of his intellectual energy in comparing cultures, political systems and countries. Yet, he had sobering words about comparison. Naturally, he referred to national histories and to comparisons among countries, but his remarks can be applied to the history of science, or indeed any type of history. He wrote:

> Comparison is not a method ...; rather, it is a discursive strategy ... One has to decide, in a given work, whether one is mainly after similarities or differences.
>
> (Anderson, 2016, p. 129)

He continued by stating that it is impossible to say whether Japan, China or Korea are basically similar or basically different. Either case can be made, depending on one's angle of vision, one's framework, or the conclusions towards which one intends to move. What about Aristotelian and modern physics? What about alchemy and chemistry? Can we establish whether they are fundamentally different or fundamentally similar enterprises? Metzger employed comparisons, although, as I shall suggest below, she also employed other tools to construct her narratives. However, her use of comparison was somewhat different from that of the philosophers mentioned above. First of all, she did not use it in order to establish absolute differences of or absolute similarities, which in turn would tell us whether history of science is discontinuous or continuous. Certainly, she observed great differences among ways of organising knowledge. She emphasised the gulf between modern ways of thinking and those that she uncovered in alchemical and early chemical texts, and as she put it, between the mental *a prioris* of different periods. She did not hesitate to present Lémery and his disciples' Cartesian chemistry as a 'revolution' (Metzger, 1969 [1923], p. 289). For her Lémery no longer even understood the analogies and sympathies that had first dominated medicine and then had 'invaded the whole field of the natural sciences' (Metzger, 1969 [1923], p. 288). Her great interest in ways of thinking appeared to align her with the philosophers who favoured discontinuity, as we shall see in the next chapters.[9] On the other hand, she constructed narratives that showed links between apparently different theories and practices. Pointing out that a clear distinction of different periods in the history of chemistry could only be arbitrary, she called the history of chemistry 'a continuous evolution' (Metzger, 1987 [1932], p. 159). Moreover, she did not believe that worldviews could tidily replace one another, and that concepts employed by a later disgraced discipline, such as alchemy, are without fail discarded forever. For instance, at the end of the seventeenth century, corpuscular theories had gained a comprehensive victory over alchemic theories based on analogies and sympathies. Metzger presented these two ways of conceiving of matter not only as opposed, but also as based on incompatible worldviews. She explained that alchemists regarded nature as having a tendency toward perfection, hence their aim to remove the obstacles to the progress of iron or copper towards the perfect metal state that is gold. By contrast, corpuscular chemists regarded nature as immutable, and therefore reduced all differences to variations of placement of the corpuscles (Metzger, 1969 [1923], p. 133). Yet, for her these two worldviews found a synthesis thanks to Becher and Stahl's theory of affinities (Metzger, 1991 [1930], p. 41). Then, the theory of affinity gradually merged with Newtonian principles (Metzger, 1974 [1930], p. 160). Metzger regarded Newton's theory of universal gravitation as employing the concept of active analogy as well, as mentioned; for her this was one of the reasons why the absorption of Newtonianism did not encounter great obstacles. The re-emergence of old concepts in new contexts, however, is not *per se* evidence of continuity, but rather of a 'natural tendency' of the human mind that can take the most diverse forms.

The very structure of her books, notably *La genèse* (Metzger, 1918) and *Les doctrines chimiques* (Metzger, 1969 [1923]), is neither linear nor constructed as a succession of incompatible mentalities. Her histories are often structured rather like variations on a theme, the latter being a conceptual system, or a worldview, or a body of practices, and often all of them together.[10] For instance, she showed the diverse, and sometimes contradictory, theories and practices that were generated on the basis of the dominant concept of active analogy, as discussed in the previous chapters. All the same, her view did not exclude in principle a certain continuity or progress. Indeed, her defence of seventeenth-century pharmacy and philosophy of metals as the past of chemistry, and early-modern diverse practices and theories about angular stones as the formation of crystallography, implies a historical continuity. This 'continuity', however, should not be understood in terms that philosophers such as Meyerson used, which meant identity of conceptual framework across the history of science. Rather, for her one or more of many 'variations' on a way of thinking and doing would be linked to following variations, possibly on a different dominant *a priori*, in the form of an enduring habit, practice or concept. Scholars' conscious opposition to a previous way of studying a particular object for her is also evidence of a historical connection. Those threads in Metzger's narrative are multiple and complex, indeed sometimes they may be messy. In many cases for her a solid historical link could be established by following the threads of authors who cited previous ones, and who accepted, amended or opposed their predecessors' ideas.

Metzger's attitude toward the question of continuity versus discontinuity of the history of science does not seem to fit with either the continuist or discontinuist views that she found in her milieu. This is because she was not interested in establishing 'absolute differences', or 'absolute similarities', in Anderson's terminology. While certainly comparing theories, and stressing differences as well as similarities, she did not mainly rely on comparison in constructing her narratives. Rather, she paid attention to the variety of small adjustments in theories, practices, attitudes and contexts that affected the research into the transformation of matter and the study of crystals. Paracelsian theories may look as contrary to current science as we can imagine. Yet, Metzger observed that Paracelsian practitioners, by rejecting the previous assumptions that astral phenomena were the model of the sub-lunar world, could then observe the human body without those constraints. Indeed, Metzger argued that 'all points of view' had become 'equivalent'; hence, 'without an apparent revolution', the last obstacles to freedom of thought had been removed. The consequences of the removal of those assumptions were not clear at the time, or for the practitioners themselves. However, the historian can observe them in the medics', pharmacists' and chemists' new theories (Metzger, 1969 [1923], pp. 153–154). In the history of crystallography, scholars who could be seen as marginal, for instance those who only focussed on one type of crystals, and the amateurs who collected and observed 'angular stones', for Metzger are part of the fabric of the history of this discipline, as they provided data and indeed made discoveries that were later used by crystallographers. Hence, for her it is not just a matter of conceptual differences, to which she nevertheless paid great attention, but also of practices, small discoveries, and introduction of tools, such as the microscope and the balance. The fabrics of the history of crystallography and the history of chemistry for her were woven with a myriad of threads, and this is why she thought that all aspects of a period are relevant to the history of science.

Unlike many of her illustrious colleagues, she was not interested in setting a date for the birth of science. As we have seen, her history of crystallography goes further back than the received view was, whereas her *La chimie* starts much later than comparable histories of chemistry.[11] However, she did not make pronouncements about the beginning of a certain discipline, beside observing that she focussed on periods in which crystallography and chemistry were not properly formed. The latter judgement was based on historical observation, namely the lack of bodies of practitioners regarding themselves as chemists or crystallographers, holding a reasonably homogeneous system of methods, theories and practices, and being bound by professional links. By contrast, philosophers and historians in her milieu compared past and present ways of thinking in order to establish the date of birth of science and, as a result, its identity. Pierre Duhem, by finding similarity between Aristotelian, medieval and current science, established that science did not start with the so-called scientific revolution, but rather had a much longer history. This continuity also erased the opposition between Christian philosophy and modern science. By contrast, Alexandre Koyré set the birth of modern science in the sixteenth and seventeenth centuries, when Galileo and then Newton replaced the closed world of old times with an infinite universe (Koyré, 1958 [1957]). Gaston Bachelard conceived of science as a rational enterprise, and its history as a history of progressive rationalisation. This view led him to set the date of birth later than most other historians of his time: the end of the eighteenth century. In the context of national history, Benedict Anderson remarked that '...longitudinal comparisons of the same country over a long stretch of time are ... important', because there is 'a vested interest in continuing and perpetuating an ancient "national identity"' (Anderson 2016, pp. 130–31). Similarly, most French philosophers and historians of science in Metzger's times founded the identity of science on its history. However, unlike national identity, its ancient origin did not add value according to most scholars. It did for Duhem, who promoted a view of science in continuity with faith. For many others, a relatively recent birth of science showed science as authentic knowledge, independent of systems of religious, mystical and magic beliefs. The birth of science went hand in hand with the age of reason. Were those historical judgements, or rather epistemological views that bent history to their ends? This is one of the questions Metzger grappled with, as we shall see in the next chapter.

## Notes

1 I shall discuss the role of philosophy and religion in Metzger's work in Chapter 5 and Chapter 6 respectively.
2 Rey distinguished his own positivism, or 'absolute positivism' (Rey, 1909a, Rey, 1909b) from Comte's positivism. He regarded his philosophy as 'scientism' (Rey, 1908, p. 6, n. 1).
3 On the point of attribution of discoveries, see Redondi in Koyré (1986).
4 On the other hand, Mieli wrote very positive reviews of Metzger's works (Mieli, 1923, Mieli, 1927, Mieli, 1930).
5 Discontinuity in history, as well as between scientific knowledge and common knowledge, has had an enduring importance in history and philosophy of science as in other disciplines. Bachelard's concept of epistemological break has been received and variously employed by such scholars as Louis Althusser, Michel Foucault and Pierre Bourdieu.
6 For a comparison between Berr, Koyré and Metzger, see Redondi (1997).
7 See Introduction and Chapter 7.
8 It is Koyré who has made the expression 'scientific revolution' familiar, with all its implications of discontinuity, and of the beginning of scientific thinking and a new way of seeing and interacting with nature.
9 Jan Golinski has compared Metzger's analysis of conceptual structures with Michel Foucault's project in *The order of things* (Foucault, 1970 [1966]). He thinks that in both cases their approaches prevent them from successfully explaining change (Golinski, 1987).
10 Metzger's *Attraction universelle et religion naturelle chez quelques commentateurs anglais de Newton* is about 'theological variations' on the 'theme of

universal attraction'; the title of Section 2 being: 'Dialectiques des variations théologiques sur le thème de l'attraction universelle', and the title of Section 3: 'Étude critique de quelques variations théologiques sur le thème de l'attraction universelle' (Metzger, 1938).

11 See Chapter 2.

# References

Anderson, B. 2016. *A life beyond boundaries: A memoir*, London, Verso.

Anonymous [HENRI BERR]. 1900. Sur notre programme. *Revue de synthèse historique*, 1, 1–8.

Anonymous. 1936. Programme détaillé du certificat d'histoire et philosophie des sciences et du diplôme de l'Institut. *Thalès*, 3, 233–247.

Bachelard, G. 1951. *L'activité rationaliste de la physique contemporaine*, Paris, Presses Universitaires de France.

Bachelard, G. 1972 [1953]. *Le matérialisme rationnel*, Paris, Presses Universitaires de France.

Bachelard, G. 1986 [1949]. *Le rationalisme appliqué*, Paris, Presses Universitaires de France.

Bachelard, G. 2013 [1931]. *Intuition of the instant*, Evanston, Illinois, Northwestern University Press.

Bachelard, G. 2000 [1936]. *The dialectic of duration, with an Introduction by Cristina Chimisso*, Manchester, Clinamen.

Bachelard, G. 2002 [1938]. *The formation of the scientific mind*, Manchester, Clinamen Press.

Bergson, H. 1998 [1907]. *Creative evolution*, Mineola, New York, Dover.

Berr, H. 1899. *L'avenir de la philosophie: Esquisse d'une synthèse des connaissances fondée sur l'histoire*, Paris, Hachette.

Berr, H. 1910. Au bout de dix ans. *Revue de synthèse historique*, 21, 1–13.

Berr, H. 1911. *La synthèse en histoire; essai critique et théorique*, Paris, Alcan.

Berr, H. 1914. La Bibliothèque de synthèse historique. *Revue de synthèse historique*, 1–3.

Berr, H. 1921. *L'histoire traditionelle et la synthèse historique*, Paris, Alcan.

Berr, H. 1950. La synthèse des connaissances et l'histoire. *Revue de synthèse*, 67, 217–238.

Berr, H. 1953. *La synthèse en histoire: son rapport avec la synthèse générale*, Paris, Albin Michel.

Braunstein, J.-F. 2006. Abel Rey et les débuts de l'Institut d'histoire des sciences et techniques (1932–1940). *In*: Bitbol, M. & Gayon, J. (eds.) *L'épistémologie française, 1830–1970*, Paris, Presses Universitaires de France.

Braunstein, J.-F. 2008. *L'histoire des sciences*, Paris, Vrin.

Brunschvicg, L. 1912. *Les étapes de la philosophie mathématique*, Paris, Alcan.

Brunschvicg, L. 1922a. *L'expérience humaine et la causalité physique*, Paris, Alcan.

Brunschvicg, L. 1922b. Letters to Sarton. *Sarton Papers, bMS Am 1803 (218)*, Cambridge, MA, Houghton Library.

Brunschvicg, L. 1923. Letter to Sarton. *Sarton Papers, bMS Am 1803 (218)*, Cambridge, MA, Houghton Library.

Brunschvicg, L. 1927. *Le progrès de la conscience dans la philosophie occidentale*, Paris, Alcan.

Brunschvicg, L. 1934. *Les âges de l'intelligence*, Paris, Alcan.

Brunschvicg, L. 1947. *L'esprit européen*, S.l., La Baconnière.

Bucciantini, M. 1987. George Sarton e Aldo Mieli: bibliografia e concezioni della scienza a confronto. *Nuncius. Annali di storia della scienza*, 2, 229–239.

Castelli Gattinara, E. 1997. L'idée de synthèse: Henri Berr et les crises du savoir dans la première moitié du XX$^e$ siècle. *In*: Biard, A., Bourel, D. & Brian, E. (eds.) *Henri Berr et la culture du XX$^e$ siècle: Histoire, science et philosophie*, Paris, Albin Michel/Centre international de synthèse.

Chimisso, C. 2008. *Writing the history of the mind: Philosophy and science in France, 1900 to 1960s*, Aldershot, Ashgate.

Duhem, P. 1991 [1914]. *The aim and structure of physical theory*, Princeton, Princeton University Press.

Febvre, L. 1952. De la *Revue de synthèse* aux *Annales*: Henri Berr ou un demi-siècle au service de l'Histoire. *Annales E.S.C.* 3, 205–218.

Foucault, M. 1970 [1966]. *The order of things: an archaeology of the human sciences*, London, Tavistock.

Foucault, M. 1972 [1969]. *The archaeology of knowledge*, London, Tavistock.

Freudenthal, G. 1990. Hélène Metzger: Éléments de biographie. *In*: Freudenthal, G. (ed.) *Études sur / Studies on Hélène Metzger*, Leiden, Brill.

Gemelli, G. 1987. Communauté intellectuelle et stratègies institutionnelles. Henri Berr et la fondation du Centre International de Synthèse. *Revue de synthèse*, IV series, 2, 225–259.

Golinski, J. V. 1987. Hélène Metzger and the interpretation of seventeenth century chemistry. *History of Science*, 25, 85–97.

Koyré, A. 1935. Au l'aurore de la science moderne. La jeunesse de Galilée (I). *Annales de l'Université de Paris*, 10, 540–551.

Koyré, A. 1936. Au l'aurore de la science moderne. La jeunesse de Galilée (II). *Annales de l'Université de Paris*, 11, 32–56.

Koyré, A. 1937. Galilée et l'expérience de Pise: à propos d'une légende. *Annales de l'Université de Paris*, 12, 442–453.

Koyré, A. 1939. *Études galiléennes*, Paris, Hermann.

Koyré, A. 1958 [1957]. *From the closed world to the infinite universe*, New York, Harper.

Koyré, A. 1968 [1943]. Galileo and Plato. *In*: *Metaphysics and measurement*, Cambridge, MA, Harvard University Press.

Koyré, A. 1978 [1939]. *Galileo studies*, Hassocks, The Harvest Press.

Koyré, A. 1986. *De la mystique à la science. Cours, conférences et documents 1922–1962, édité par Pietro Redondi*, Paris, Ed. de l'École des hautes études.

Metzger, H. 1918. *La genèse de la science des cristaux*, Paris, Alcan.

Metzger, H. 1937. Pierre Duhem, la théorie physique et l'histoire des sciences. *Archeion*, 19, 135–139.

Metzger, H. 1938. *Attraction universelle et religion naturelle chez quelques commentateurs anglais de Newton*, Paris, Hermann.

Metzger, H. 1969 [1923]. *Les doctrines chimiques en France du début du XVII$^e$ à la fin du XVIII$^e$ siècle*, Paris, Blanchard.

Metzger, H. 1974 [1930]. *Newton, Stahl, Boerhaave et la doctrine chimique*, Paris, Blanchard.

Metzger, H. 1987 [1932]. Eugène Chevreul historien de la chimie. *In*: Metzger, H. 1987. *La méthode philosophique en histoire des sciences. Textes 1914–1939, réunis par Gad Freudenthal*, Paris, Fayard.

Metzger, H. 1987 [1935]. La littérature chimique française aux XVIIᵉ et XVIIIᵉ siècles. *In*: Metzger, H. 1987. *La méthode philosophique en histoire des sciences. Textes 1914–1939, réunis par Gad Freudenthal*, Paris, Fayard.

Metzger, H. 1991 [1930]. *Chemistry*, West Cornwall, CT, Lucust Hill Press.

Metzger, H. 2019 [1937]. The philosophical method in the history of science. (Appendix 4).

Mieli, A. 1916. Sul concetto di storia della scienza. *Rivista di storia critica delle scienze mediche e naturali*, 7, 42–46.

Mieli, A. 1919. Bibliografia metodica dei lavori di storia della scienza pubblicati in Italia. *Archivio di storia della scienza*, 1, 84–86.

Mieli, A. 1923. Hélène Metzger, *La genèse de la sciences des cristaux*; *Les doctrines chimiques en France du début du XVII à la fin du XVII siècle*. *Archivio di storia della scienza*, 4, 290–291.

Mieli, A. 1927. Hélène Metzger, *Les concepts scientifiques*. *Archeion*, 8, 270–271.

Mieli, A. 1930. Hélène Metzger, *La chimie*, Paris 1930 (et. al). *Archeion*, 391–394.

Mieli, A. & Brunet, P. 1935. *Histoire des sciences. Antiquité*, Paris, Payot.

Redondi, P. 1997. Henri Berr, Hélène Metzger et Alexandre Koyré: la religion d'Henri Berr. *In*: Biard, A., Bourel, D. & Brian, E. (eds.) *Henri Berr et la culture du XXᵉ siècle. Histoire, science et philosophie*, Paris, Albin Michel/Centre international de synthèse.

Rey, A. 1908. *La philosophie moderne*, Paris, Flammarion.

Rey, A. 1909a. Sur le positivisme absolu. *Revue philosophique*, 34, 65–66.

Rey, A. 1909b. Vers le positivisme absolu. *Revue philosophique*, 34, 461–479.

Rey, A. 1930. Histoire de la science ou histoire des sciences. *Archeion*, 12, 1–4.

Section d'histoire des sciences. 1936. Séance du 22 janvier 1936. *Archeion*, 18, 238–247.

# 5 History of science and philosophy

> ... if one wanted to do history of science seriously, and find a public of readers, it was necessary to look among philosophers, historians, and even professors of literature
>
> (Metzger, Letter to Sarton, 20.6.1922)

## The roles of philosophy in Metzger's history of science

It is not surprising that in the letter to Sarton quoted above, Metzger should have mentioned philosophers first in the list of intellectuals who did history of science seriously. This reflected her own experience, as well as the institutional position of her discipline in France. It is through her frequentation of philosophers and attendance of philosophy lectures at the Sorbonne that she found her first congenial milieu, as I sketched in the Introduction. Her increased knowledge of philosophy and philosophical approaches to the history of science enabled Metzger further to developed views and questions that were already present in her first book. In her work, philosophy appears to be relevant in different ways. First of all, philosophy is an object of her study. It goes without saying that as far as her period of choice is concerned, a distinction between the study of nature and philosophy could not be drawn, certainly not in the manner in which it is drawn in modern times. Moreover, for her, in order to understand the questions that past scholars aimed to answer, it is necessary to know their metaphysics, this being both the metaphysics that scholars unconsciously held, and, for some of them, an object of study. For instance, in the eighteenth century, a central question was whether light was a substance or an accident (Metzger, 1974 [1930], pp. 8–9). Moreover, scholars' methods depended on general philosophy: this is why Cartesian crystallographers set out to rely solely on common sense and deduction (Metzger, 1918, p. 124). Her aim therefore was first of all to understand past scholars' metaphysics, and the philosophy to which they subscribed. This was particularly important when there was no widely accepted theory, and indeed when practitioners had no precise theory, as in the case of her eighteenth-century experimental chemists (Metzger, 1969 [1923], p. 343). Those chemists might not have had a clear theory, but they did hold a metaphysics that guided their actions in the laboratory, as well as their gaze when observing their results. Metzger deemed the knowledge of philosophy and theology to be crucial in order to understand the texts that she studied. For her, sciences in formation are particularly shaped by philosophical systems as they lack their own structure, methods and disciplinary identity. In her first book, crystallography appeared to her to be particularly at the mercy, as it were, of philosophy. She remarked that eighteenth-century crystallography was nothing more than a development of philosophical systems (Metzger, 1918, p. 212). She ironically wrote that it was a 'superb monument', which was all very solid, with the exception of its tenuous empirical basis – no more than an 'indication' (Metzger, 1918, p. 213). On the other hand, she presented Stahlian chemistry on its way to gaining autonomy, navigating as it did between empiricism and Cartesian rationalism, and striking a balance between relying on experience and intelligently interpreting it (Metzger, 1918, pp. 148–149). With time, the sciences acquired independence from philosophical systems, and developed their own methods.

However, independence from the discipline of philosophy in a strict sense does not mean independence from metaphysical assumptions and particular ways of reasoning. For Metzger, no science, or any other discipline, can exist without them, as they are products of our minds. The historian of science needs to understand scientists' minds. As a consequence, philosophy remains important for her no matter what the stage of the science under study is. She also advocated a 'philosophical method' of the history of science. But what type of philosophy should guide the historian of science? Her views of scientific and historical knowledge led her to reject empiricism both as a general philosophy and as a historical method. As we shall see, she articulated her opposition to the former in her criticism of the Vienna Circle's philosophy, and to the latter in her rejection of the Centre de synthèse's official 'objective' method. She did not accept that her sources could be studied as an 'inert object' (Metzger, 2019 [1937]). Rather, her aim was the study of the minds behind theories and practices, which could be reached using a variety of tools, including, as seen in Chapter 3, sympathy. Her engagement as a historian with her sources led her to develop a reflective attitude towards her own understanding of texts. As a historian, she did not regard herself as a recorder of facts, but rather as another subject, with particular knowledge, worldview, conceptual framework and emotions. All this for her had to be considered in order to represent past texts as faithfully as possible.

She also regarded history of science as having a philosophical aim, namely to understand the human mind. Moreover, she thought that the knowledge so acquired would in turn help us improve our reasoning, in philosophy as well as in science. Philosophy, therefore, for her is both an object of study and the main aim of the history of science. Was history of science then to

replace philosophy? Metzger appears implicitly to argue against this proposal when she emphasised the division of labour between the historian and
the philosopher. Often, it seems that for her history of science must provide
the material on which the philosopher can work and draw her philosophical conclusions. At other times, one has the impression that for her history
of science is able, on its own strength, to provide philosophical answers.
Indeed, she developed her own historical epistemology. Did she think that
her epistemology emerged from her own study of the histories of chemistry
and crystallography? Or was it rather the result of her 'mental *a prioris*'
combined with her study of history, perhaps similarly to the work of scientists, who in her view have their own presuppositions and mentalities, which
shape their phenomena? No clear answer to these questions can be found in
her writings, and arguably no answer could be given, because history and
philosophy for her shape and support each other. Her attitude put her in
a distinctive position in her intellectual milieu. Philosophers such as Émile
Meyerson, Léon Brunschvicg and Gaston Bachelard employed history of
science in order to elaborate their epistemologies. This made her close to
them, but also critical of their use of history, which for her was often rather
superficial and narrowly instrumental. On the other hand, she admired
the scholarship of historians, for instance her long-lasting correspondent,
George Sarton. At the same time, history without a philosophical approach
and aim seemed not only a little arid to her, but also prone to be historically
less accurate than it could have been. She was particularly critical of those
scholars who thought that past scientific texts could be read without considering their philosophical contents and assumptions. Similarly, she decisively
rejected the view, shared by such diverse scholars as Meyerson and Berr,
that historians' philosophical assumptions could be brackcted in their interpretation of past texts.

## The role of the *a priori* and her rejection of empiricism

Metzger chose to present her historical method as 'philosophical' at a talk
at the Centre de synthèse. Her polemical intention is clear, and arguably
excessively shown in her opening remarks (Metzger, 2019 [1937]). A few
seconds into her talk, she imagined the reaction of her audience to be one
of puzzlement over the very idea of uniting history and philosophy. She had
been invited to give this talk by the director of the Centre, Henri Berr, and
it is apparent that she was contrasting her own historiographical views with
his. We have seen in Chapter 4 that Berr, as well as Abel Rey, promoted an
empirical method in history, which made their view of historical synthesis
very different from Metzger's defence of the importance of the study of
culture and society to understand past texts. In fact, the philosophical differences between Metzger and Berr had a more general impact on their disagreement about historical and indeed scientific knowledge. Berr aimed to
make history into a positive science and proposed his historical synthesis as

a replacement for philosophy of history (Berr, 1911, pp. 22–23). He claimed that he did not reject philosophy overall, but only traditional philosophy. Indeed, he approved of current philosophy based on 'scientific knowledge' (Berr in (Metzger et al., 1935)). By contrast, Metzger, while defending the role of philosophy in the history of science, opposed empiricism and scientific philosophy. Her views, and indeed emotions, on empiricism and scientific philosophy are very clearly exemplified in her brief account of the 'First International Congress of Scientific Philosophy'[1], in which she attacked the Vienna Circle's 'total empiricism' (Metzger, 1987 [1935]-b, p. 166). She regarded logical empiricism as contrary to her own view of science, philosophy and history. As she put it elsewhere, Vienna Circle philosophers declared that they would 'succeed in solving all the problems to be solved by relying on protocol statements or reports of their laboratory work' (Metzger, 2019 [1936]). For her, these 'unruly' philosophers did not grasp the nature of knowledge, which always needs the contribution of our mental categories. She could not approve of what she saw as the Vienna Circle's attempt to separate 'thought' from the 'thinker', and their conception of thought as a 'thing' (Metzger, 1987 [1935]-b, p. 167). She thought that they aimed to imprison knowledge in strictures that would impede its progress. Indeed, she wondered whether 'the Vienna Circle has not first killed within itself the expansive thought that remains a source of inspiration, even when it is educated and disciplined by reflective thought'. In fact, for Metzger not even reflective thought, which we can loosely identify with the 'context of justification', can thrive under the logical empiricist's regime. Indeed, she also wondered whether 'the Vienna Circle has not killed within itself the reflective thought which, on every occasion of doubt, goes back to its nascent state to pronounce a judgment'. In other words, for Metzger there is no clear line between 'context of discovery' and 'context of justification'. Expansive thought in fact helps reflective thought when the latter cannot solve problems that require creativity and new solutions. The Vienna Circle stopped all this with what for her was a return to 'the most barbaric of scholastics' (Metzger, 2019 [1936]). Moreover, she could not accept what she saw as the Vienna Circle's view of history of science and philosophy, namely as having a conclusion, when methods and language would be established once and for all. For her, just as for Brunschvicg and Bachelard, the history of science has not ended, and neither has history of philosophy; the mind continues its quest, creating new ways of approaching nature.

She depicted the Vienna Circle's members as 'scientists who have discovered the attraction of philosophy' (Metzger, 1987 [1935]-b), an interesting remark for somebody whose academic qualifications were in science. Scornfully, she wrote that the 'classic philosophy' that they despised was in fact the philosophy taught in German high schools. Their 'systematic ignorance' of history of philosophy led them wrongly to believe that their views were new. She regarded Carnap's project of creation of a perfect language as an enterprise that had been repeatedly attempted in the history of

philosophy, by Condillac, Leibniz, Pascal and Ramon Llull. She emphasised that this did not mean that the project was unworthy, but rather that
Carnap and his associates supposed that it had never been pursued previously; in short, that they did not know history. Her tone betrays that she
was opposed to the Vienna Circle's aims not only epistemologically, but
also ethically, or politically in a broad sense. She presented their philosophy
as fundamentally illiberal, although she would not have used this term. She
even joked that they would get a universal agreement by imprisoning or
annihilating those whom they could not convince. Her flippant remarks,
which of course she did not mean literally, with hindsight may seem tragi-
cally wrong, as in fact many members of the Vienna Circle were the target of
antisemitism. Her remarks came from a brief, although published, account
of a conference, and she did not spend much time studying logical empiricism. As a consequence, her judgements may appear at times uncharitable.
However, they do illustrate her strongly held epistemological views. For
her, it is precisely the study of history that shows the prominent role of the
mental *a priori* in the construction of scientific theories.

The mental *a priori* for her plays a role in any type of knowledge, including historical knowledge. Historians who proposed an empiricist view of
history as, in her presentation, an accumulation of facts without interpretation or theory, wrongly interpreted their own practice. In her words:

> Some slightly naïve historians thus could have believed that the chrono-
> logical accumulation of scientific texts would reveal the true progres-
> sion of our intelligence directly by itself, excluding all interpretation,
> criticism or commentary. They asked this overwhelming body of texts
> to reveal the truth by silencing once and for all philosophers, who were
> invited to stop taking up public space with the deafening clamours of
> their pointless disputes. With this objective achieved, [the historians]
> offered unlimited thanks to history that had delivered all its secrets to
> them.
>
> (Metzger, 2019 [1935])

One such 'slightly naïve' character was the historian of chemistry Maurice
Delacre. Metzger commented that he thought that history was a collection of objective facts without theory or philosophy because, by his own
admission, he was already an empiricist before turning to history. 'If he had
been a mechanicist, an atomist, or a theorist', she remarked, 'the response
that he would have received [from his sources] would have perhaps been
something else entirely'. Empiricists, for her, in history just as in science,
are in fact naïve as they do not realise that their own empiricism is still
a philosophy through which they interpret facts and construct narratives.
She must have thought the same regarding Berr, who promoted a type of
objectivity that eliminated subjectivity and minimised the contribution of
the mind to knowledge. He regarded imagination, intuition and empathy

as particularly dangerous. Similarly, he criticised the role of *a priori* knowledge, and defended the centrality of *a posteriori* knowledge. In this, he was in agreement with his associate Lucien Febvre, the historian founder of the journal *Les Annales*, and co-director of the Centre de synthèse. In the entry on 'History' that they wrote together for the *MacMillan Encyclopaedia of the Social Sciences*, they presented history as recreated – as opposed to created – by the mind. The historian produces an image of the historical past, which exists thanks to this 'recreation'. However, they warn against the mind's 'creations', and in particular against 'the aesthetic pleasure derived from the playing of imagination and sensibility on the past, which explains the great number of easy and dangerous compromises between art and history' (Berr and Febvre, 1932, p. 357). Berr's particular target was philosophy of history. For him, nineteenth-century philosophy of history had constructed arbitrary systems, daring generalisations and downright fantasies (Berr, 1911 #312, pp. 22–23); for this reason, there could be no compromises with the German historical school, including Droysen (Berr, 1911, p. 243). We have seen in Chapter 3 that several commentators have found some similarities between Metzger's view of history and that school. This holds, with all due differences, especially for the role that she assigned to the historian's sympathy and intuition. Berr mocked the role attributed by intuition: '*[i]ntuition*, living intuition, here is the grand word' (Berr, 1911, p. 233). He allowed for some 'subjective' elements in historical research, but those were nothing more than an open attitude: in his words, 'curiosity for all that is human' (Berr, 1911, p. 253). In fact, Metzger's view that historians have their own mental *a prioris*, which differ from those of past scholars, appeared to be almost incomprehensible to Berr and other members of the audience of her talks.[2] Indeed, Berr suggested to her that the historian's *a priori* is just lack of rigour, and that it boils down to making claims about the past without sufficient evidence (Berr in (Metzger et al., 1936)). Another member of the audience proposed a psychology or psychoanalysis of the *a priori* (Laignel-Lavastine in (Metzger et al., 1936)). In other words, they interpreted Metzger's *a priori*, when attributed to the historian, either as a tendency to formulate unfounded claims, or as the interference of emotions in the rationality of the task of history.[3]

As discussed in Chapter 3, Metzger thought that scientific and historical knowledge were similar. However, her epistemology was radically different from Berr's. For her, objectivity in Berr's sense of absence of human thought in the ordering of data is a chimera. The role of mental *a prioris* is central in science as well as in history. In her words, 'the history of science, just as laboratory observations, must be interpreted by our minds ...' (Metzger, 1987 [1929], p. 106). Similarly, talking in Berr's presence, she stated her rejection of empiricism, while emphasising the parallel between science and history:

Collecting documents is not history, and collecting measurements is not science, even if history cannot do without documents and science without measurements. History is nothing without the historian's intelligence, and physics is nothing without the physicist's intelligence. We should assert once more and without fearing the contradiction of the supporters of pure experience, that the facts that history unearths cannot reveal the path followed by the human mind by their mere assemblage.

(Metzger, 2019 [1935])

She further clarified:

Just as the scientist can give various interpretations of the facts that he has decided to study, likewise the historian of science can give diverse interpretations of the texts that he has decided to use in his construction.

(Metzger, 2019 [1937])

Moreover, Metzger did not think that the removal of subjective elements, such as intuitions and emotions, would render history more similar to the sciences. Her historical research, and presumably her scientific studies at university level, showed that chemists and crystallographers employed their intuitions and emotions in their research. In short, the disagreement between Metzger and Berr was not so much about whether history was a science, but rather about knowledge, both scientific and historical. Their differences were philosophical, and in particular epistemological. Metzger aimed to emphasise the role of the *a priori* in the construction of historical narratives as well as scientific theories. Although ostensibly reporting Federigo Enriques' view, she expressed her own when she wrote that:

history is no more an assemblage of data than science is an assemblage of facts. The history of science, just as science itself, is a construction of the mind. The historian is not a simple collector of documents; he constantly exercises his faculties as a creator and as a critic. He works on texts just as the scientist works on facts brought about in the laboratory.

(Metzger, 1987 [1934], p. 148).

Metzger did not say that Enriques criticised empiricism in the history of science in the context of his attack on Duhem's history of science.[4] Metzger did not criticise Duhem in her work. However, she enlisted Enriques in her anti-empiricist battle, inviting those who opposed the *a priori* to reflect on 'Enriques' solid and brilliant' argument (Metzger, 1987 [1934], p. 149). It is true that many of Enriques' ideas, though not all,[5] look very close to Metzger's. His opposition to positivism, empiricism, as well as to idealism and pragmatism must have resonated with her. He gestured substantially towards a historical neo-Kantianism that was very close not only to Metzger's in particular, but also to that of important philosophers in France, including Léon Brunschvicg. Metzger had partly absorbed the opposition to empiricism, as well as a historical neo-Kantianism, from her informal studies at the Sorbonne, but she had also derived these positions from her own understanding of science through her historical work.

She characterised her method as 'philosophical', not only because of the importance of the study of the minds behind the works, but also, and crucially, because of the engagement of the historian's mind in the construction of her narrative. For her, philosophy was not opposed to science, just as history was not opposed to it. Needless to say, she rejected specific philosophies, notably empiricism, as we have just seen. More generally, she opposed philosophies that would not allow for historical variations in the use of concepts; indeed, she believed that the study of the history of science would definitely contradict them. In her words:

> [I]t seems clear that the study of the history of science would eventually heal the philosopher (if the disease could be cured) of the strange habit of wanting to present definitive *a priori* or *a posteriori* concepts, on which the mind could quench its thirst for certainty, and which one could rightly call concepts of divine right.
>
> (Metzger, 2019 [1935]).

These words made her close to Brunschvicg, Bachelard, Koyré, and to all those philosophers we have come to call historical epistemologists. Yet, as we shall see in the next section, she also had words of criticism for the philosophers who, like herself, regarded the mind as historical, indeed sometimes more thoroughly so than she herself did.

## The role of history and her criticism of *a priori* philosophy of science

The philosophers in Metzger's milieu paid great attention to the history of science, and the development of history of science as an institutional discipline was dominated by philosophers. Both the philosophical attention to history of science and philosophers' promotion of this discipline had started in the nineteenth century. Famously, Comte himself promoted the first chair of history of science at the Collège de France. Positivism, however, was a very complex legacy for the philosophers with whom Metzger worked. It was fundamental for Abel Rey,[6] but for other philosophers in Metzger's milieu, Comte's view of history of science was a critical target rather than a model. Most philosophers regarded the history of science as a history of progress, or, at least, regarded current science as superior to past science. However, the type of progress that they thought they could detect in the history of science exhibited fundamental differences from Comte's. A case in point is Brunschvicg's philosophical histories (Brunschvicg, 1912); (Brunschvicg, 1922); (Brunschvicg, 1934), which, as mentioned in Chapter 4, display a type of progress which is not linear and does not obey laws, against Comte's model. Brunschvicg certainly had norms by which he judged progress, namely rationalisation, secularisation and mathematisation of knowledge, as I discussed elsewhere (Chimisso, 2008). These very norms, though, for him showed that there were setbacks in the progress of science and thought; indeed, that progress could stop or even reverse. For him, there is no necessity in scientific development. Importantly, for Brunschvicg, just as for Bachelard, Canguilhem and historical epistemologists in general, there is no predetermined end to the history of science and the history of the human mind. Brunschvicg compared Comte's philosophy of history to that of Hegel and Spencer, and declared all of them 'happily' out of date (Brunschvicg, 1936, p. 30). There is no doubt that Metzger was among the opponents of the view of history as an ordered and closed process that positivist philosophy proposed. Positivist historiography for her is *a priori*, and ignores the great lessons that serious engagement with history delivers. She regarded Comte only as a philosopher, who had little understanding of history. She remarked that his mind 'was not built for the patient and meticulous study of history', but he had nevertheless wanted his philosophy to be supported by the authority of facts. Metzger rhetorically asked how facts could have failed to support his philosophy, 'since the truth that he held had been conclusively established' (Metzger, 2019 [1935]). In her reading, far from extracting his law of three stages from the study of history, Comte had formulated his law *a priori*, and then selected historical facts that fitted with his theory. She reflected that '[b]y discovering the value of the history of science and showing it to the world, the victorious positivism had hoped to enslave it once and for all' (Metzger, 2019 [1935]). Although she described her method and the aim of history of science as philosophical, for her history of science must not be enslaved to philosophy. In fact, for her, history of science, if it is to be of any use to philosophy, must be proper history, rather than a self-serving collection of 'facts'.

Had philosophers learned to take history of science seriously since Comte? Metzger was more positive about her contemporaries, but she still saw some distance between those philosophers' attitude to the history of science and her own. She welcomed Brunschvicg's view of history of science, especially in its opposition to Comte's. In particular, she stressed that Brunschvicg had avoided nineteenth-century *a priori* schemes, which presented the supposed deterministic mechanisms of nature and mind (Metzger, 1987 [1935]-a, p. 173). Indeed, when she wrote that Brunschvicg recommended a history that it is 'contemporary of its object', she appeared to echo her own aspiration of making herself a contemporary of her sources (Metzger, 1987 [1935]-a, p. 173). Metzger also approved of his view of the mind as changeable. Although Comte too had rejected the fixity of the mind, he had plotted the evolution of the mind to its predetermined conclusion. Like many other philosophers of his time, Brunschvicg was particularly concerned with the great scientific revolutions of the late nineteenth century and of the twentieth century, namely non-Euclidean geometries and the theory of relativity. These revolutions showed to him that Kantianism must be regarded as the philosophy of Newtonianism; twentieth-century science called for an updated critical rationalism. No epistemology can be conceived once and for all, independently of the advancement of science. For Brunschvicg it is the history of science that shows that the mind is not fixed (Brunschvicg et al., 1923, pp. 147–148) (Brunschvicg, 1922). Metzger appreciated Brunschvicg's approach, which must have appeared to her as a validation of her own reading of past texts. Her 'strange' primary sources were then expression of ways of thinking that differed from those behind current science; this was because the mind had a history, and this history was neither determined in advance, nor linear. She commended Brunschvicg's philosophy for showing 'the sinuosity, detours and progressive adaptation' of thought (Metzger, 1987 [1923], p. 172).

Yet, his relativism had the limits to be expected of an epistemologist who regarded modern science as the pinnacle, however provisional, of human thought. Unsurprisingly, she was troubled by his epistemological approach, which judged past inquiry into nature as either science or non-science. She seemed unwilling to oppose Brunschvicg completely, and conceded that perhaps Aristotle and Bacon could be excluded from the history of science. However, she could not accept his rejection of pre-Lavoisierian chemists. She retorted that, if the latter do not belong to the history of science, then neither do Boyle, Lémery, Stahl and Boerhaave, as none of them followed experimental practices that modern science would accept as valid (Metzger, 1987 [1935]-a, pp. 173–174).

The divergence between Metzger's approach and that of philosophers close to her becomes clearer when we read her evaluation of Brunschvicg's pupil, Gaston Bachelard. She expressed great admiration for him as a philosopher and praised his pedagogical intents.[7] However, while acknowledging his erudition, she could not accept his use of history. In her review of *La formation de l'esprit scientifique* (Bachelard, 1993 [1938], Bachelard, 2002 [1938]), she distinguished between Bachelard the philosopher and Bachelard the historian (Metzger, 1987 [1938], p. 190). She found the latter wanting. Her criticism of Bachelard as a historian is very much in line with her view that philosophers often selected historical examples purely to serve their philosophical aims. In *La formation*, Bachelard aimed to show that there was an epistemological break between what he called pre-science and modern science. The latter for him only started at the end of the eighteenth century, with the age of the scientific mind. The previous mentality, which he called the 'pre-scientific mind' did not produce any science, but only projections of its own emotions and desires. In short, he counterposed an imaginative and a rational approach to the study of nature, with the former being an obstacle to the latter. Since for him science emerges in opposition to pre-science, early-modern natural philosophy for him was just a repository of the works of imagination. The works of the imagination, it is important to add, for Bachelard are not part of a historical development at all, as for him human imagination and desires are fixed. In fact, he excluded early-modern natural philosophy not only from the history of science, but indeed from history altogether, as for him it can be studied in the same way as one analyses modern poetry. By contrast, scientific theories and practices are the expression of reason, which changes and advances; this is why one can write a history of science. How did Bachelard gather his evidence to support his claim that early-modern natural philosophy was simply a work of the imagination, similarly to poetry and art? According to Metzger, he did not do it very well. She observed that Bachelard's use of history was not 'very precise' (Metzger, 1987 [1938], p. 194). She was less severe on Bachelard than she had been on Comte, but her criticism is similar, pointing at a self-serving choice of sources. In her words, Bachelard 'is right in all that he says, and wrong about what he does not say' (Metzger, 1987 [1938], p. 194). She explained that he did not mention some important scholars, and of other scholars he only mentioned the more controversial ideas, while he was silent about their important contributions. For her, he had selected what we could call soft targets. For instance, she pointed out that many of the scholars on whom Bachelard focussed were not taken seriously even in their own times. She wrote, 'crystallographers of value such as Romé de l'Isle had been furious at Robinet's popular [*mondaine*] success' (Metzger, 1987 [1938], p. 195). Nevertheless, Bachelard used Robinet's work extensively in order to show how the pre-scientific mind works. He cited Robinet's work, or, in his words, his 'pedantic reverie', to illustrate several epistemological obstacles to scientific knowledge, including pragmatic knowledge, the animist obstacle, the myth of digestion, avarice as the basis of realism, and sexual desire, all of which misguide pre-scientific minds in their interpretation of nature (Bachelard, 2002 [1938], pp. 101, 146, 163–164, 180, 192–193). It is clear that for Metzger a more equitable approach to sources would have produced a history of science exhibiting a very different shape from Bachelard's, and a less easily drawn demarcation between science and pre-science. In particular, and almost in passing, Metzger observed that the 'light' of modern science, when projected on the past, may make past texts look far more obscure than they really are (Metzger, 1987 [1938], p. 194). In other words, she saw in Bachelard's history of science a problem that mirrored the 'triumphalist' histories of science that she took as her critical targets. Whereas the latter projected current theories and discoveries onto the past, and there found supposed anticipations, Bachelard, judging the past on the basis of modern science, only saw differences, and failed to grasp the achievements of early-modern natural philosophy. Metzger wondered whether future scientists would not regard current science as partially pre-scientific (Metzger, 1987 [1938], p. 195).

In *La formation*, Bachelard had judged alchemy and early chemistry, i.e. Metzger's sources, as separated from science by an epistemological break. A break for him not only occurred historically, sometime at the end of the eighteenth century, when the history of science started. It also occurred synchronically, separating science from non-rational activities. In fact, what

Metzger regarded as her object of study for Bachelard was the opposite of science. As we have seen in Chapters 2 and 4, Metzger aimed to study the average opinion [*opinion moyenne*] of professionals and amateurs alike, in a given time and place. Her '*opinion moyenne*' was very close to the *bête noir* of Brunschvicg and Bachelard's historical epistemology, namely '*sens commun*', or 'common sense', with a negative connotation. Common sense must not be confused with '*bon sens*', although the latter is often translated with the former in English. *Bon sens* means 'good sense or reason' [*le bon sens ou la raison*], as Descartes, and his translator, put it (Descartes, 1985 [1637], p. 111). Most philosophers at the time understood 'common sense' as Lalande's *Vocabulaire technique et critique de la philosophie* defined it:

> In the current philosophical language, common sense [*sens commun*] is the set of opinions so generally accepted at a particular time and in a given milieu that contrary opinions appear as individual aberrations, which would be superfluous to reject seriously ...
>
> (Lalande, 1999 [1926], p. 972)

Brunschvicg attacked 'common sense', or 'universal agreement', as unable to produce scientific knowledge (Brunschvicg, 1922, pp. 572ff). Bachelard radicalised his teacher's view, as he thought that '[o]pinion *thinks* badly; it does not *think* but instead *translates* needs into knowledge ... nothing can be founded on opinion: we must start by destroying it' (Bachelard, 2002 [1938], p. 14). Common sense for Bachelard produces 'common knowledge' which is separated by an epistemological break from scientific knowledge (Bachelard, 1986 [1949], Chapter 4); (Bachelard, 1972 [1953], Conclusion). This epistemological break between common knowledge and scientific knowledge translates into historical discontinuities, as scientific knowledge can only emerge in opposition to previous knowledge (Bachelard, 2002 [1938]); (Bachelard, 1968 [1940]).

Not all philosophers close to Metzger regarded common sense and scientific rationality as opposed. She was familiar with Meyerson's opposite view. In his words:

> while it is true that what we call concepts of common sense are fashioned by an unconscious process, the process is otherwise strictly analogous to the operation by which we form scientific theories; ... here, too, the causal tendency, the principle of identity in time, plays a preponderant role, ... from this point of view common sense is an integral part of science ... ; or, inversely, that science is ... only a prolongation of common sense.
>
> (Meyerson, 1930 [1908])

Incidentally, Meyerson's concept of 'common sense' does not have the social dimension of Lalande's definition, but rather indicates the spontaneous approach to reality that is common to all human beings. Does this mean that Meyerson's claim that 'common sense is an integral part of science' corresponds to Metzger's view that spontaneous thought is part of science? It does not. Metzger did not hold that thought has one and only particular structure, and she was not really interested in the unconscious mechanisms of thought, but rather in how various concepts play different roles in different times and places. Moreover, whereas Meyerson saw the operation of 'common sense' and logic as exhibiting the same structure, Metzger counterposed 'spontaneous thought' to 'reflective thought', both necessary to science, but different. 'Spontaneous thought' for her is freer and often less logical than reflective thought, and may take different and contradictory paths. Metzger did not engage directly with comparisons between her view of the mind and Meyerson's. As she did with Bachelard's, she preferred to read Meyerson's work from the point of view of a historian of science. However, she used rather different tones in her comments on the books of these two philosophers. Her papers on Meyerson are a little awkward to read, punctuated as they are by declarations of admiration closely followed by attacks that betray resentfulness on her part. She often remarked on Meyerson's knowledge of history of science, and wrote that he was a historian before becoming a philosopher (Metzger, 1987 [1929], p. 106). Nevertheless, she classed him as a philosopher, though in the group of philosophers whose work was based on the history of science, including Comte, Cournot, Renouvier, Mach, Duhem, Rey and Brunschvicg. All the same, she emphasised that his epistemological questions differed considerably from hers, and that his work was not consecrated to the history of science (Metzger, 1987 [1929], p. 95). She appeared to avoid discussing Meyerson's epistemology directly. She even wrote that she would not contest Meyerson's view that all scientific theories progress by reducing diversity to unity, but that on the other hand she focussed on the 'theory of the formation of scientific concepts', which for her was not in the same domain as Meyerson's 'psychological and logical analysis'. She rather aimed at understanding the differences between the different 'systems of concepts posed by our intelligence' (Metzger, 1926, p. 154). In fact, Metzger was much more sceptical about Meyerson's epistemology that she let on in *Les concepts scientifiques*, and I think she perceived it as a threat to her own view of science. Still conceding a little, she asserted the variability of mentalities with the following words:

> [E]ven if the human mind is always and everywhere similar in its fundamental characteristics, even if it truly has an unalterable framework, the attitudes that it is able to adopt and that effectively determine the orientation of men's mentality are diverse and highly variable.
>
> (Metzger, 2019 [1937])

Metzger had spent a great deal of effort to show the distinct concepts behind the astonishingly different scientific theories that she found in her sources.

Meyerson seemed to belittle such efforts by finding a supposedly deeper explanation, namely that identification is at the bottom of all concepts and all reasoning, and by therefore asserting that there is no fundamental difference between the respective ways of reasoning of early-modern natural philosophers and of modern scientists (Meyerson et al., 1936); (Meyerson, 1930 [1908]).

One of Metzger's objections was that Meyerson was selective in what he accepted as evidence. For instance, she quoted a passage where he argued that the last lines of Newton's *Principia* should not be taken seriously because they echoed the philosopher Henry More. Metzger retorted that in fact, as 'Burtt and Snow have demonstrated, Henry More's philosophy played an important role in the formation of Newton's physics' (Metzger, 1987 [1936], p. 112). For her, the metaphysics behind a scientific theory is crucial, but philosophers like Meyerson may choose to ignore it when it does not fit their view of science. On the other hand, Meyerson for her found in the history of science what he had postulated from the beginning; in other words, she accused him of begging the question. She wrote that he had warned her not to introduce into her histories ideas that would have been alien to the authors of her sources, and that he himself aimed to stay within the limits of common sense. Indeed, she ironically commented, his whole construction is plausible if one admits with him that science has always been based on common sense's vague philosophy. In this supposition, she added, the 'epistemology of identification' is simply justified (Metzger, 1987 [1936], p. 111). Here Metzger referred to Meyerson's view that 'identity is the eternal framework of our mind' (Meyerson, 1930 [1908], p. 284): in common sense as in science, for him the mind exhibits the tendency to identification. We perceive differences in our experience of the world, and we aim to eliminate them. He famously interpreted chemical formulas (his example is $Na + Cl = NaCl$) as partial identifications that do not differ in form from the thought processes of native Australians, who, in Lévy-Bruhl's description, partially identify with parrots (Meyerson, 1931, pp. 81–88). In other words, he regarded all ways of reasoning, in mathematics, in empirical sciences, in common sense, in magic and religion, as formally the same. Metzger repeated her judgement of his epistemology elsewhere, by noting that Meyerson, by staying within the limits of common sense, 'marvellously' managed to describe all thought in terms of common sense (Metzger, 1938, p. 24).[8]

While wholeheartedly approving of philosophers' interest in the history of science, Metzger objected to what she saw as their *a priori* attitude towards history. Their theories appeared to her to precede their study of history of science, which then they selectively or superficially used. Moreover, her view of the relationship between everyday concepts and theories on the one hand and science on the other did not align either with Bachelard's or with Meyerson's. She studied the very significant differences in thinking behind various inquiries into nature. Only by accepting this variety, and rejecting Meyerson's reductionism, could she make sense of her sources. However, unlike Bachelard, she saw no epistemological break between past inquiries and modern science, nor did she think that science is fully rational. She defended the role of ways of thinking that differ from the fully logical and ordered manner that appeared to be implied by the finished products of modern science. For her, those polished products hid the process of creation, which involved a great deal of 'spontaneous thought', of creativity and emotions.

## The philosophical aims of history of science

As we have seen, Metzger strongly opposed empiricism, both as a general philosophy and as a historical method. For her, the contribution of the mind to the construction of scientific knowledge as well as to the construction of historical narratives was fundamental. Just as chemists need to experiment, so for her the historians' minds need to interact with their sources. As cited, she remarked that the historian's mind is her only 'reagent'. On the other hand, she also condemned the *a priori* attitude of philosophers who extracted self-serving examples from the history of science and overlooked all the nuances, variations and complexities of historical sources. By extending her parallel between chemistry and history, one could say that for her those philosophers formulated theories without carrying out the necessary experiments, and without considering the full evidence. However, we should not think that she was equidistant between positivists on the one hand, and philosophers like Brunschvicg and Bachelard on the other. She wholly rejected positivism[9] and empiricism, and spent much of her energy in asserting the importance of the role of concepts in the organisation of data. Her study of scientific concepts and their different roles, such as the dominance of active analogy in Renaissance thought, made her a true historical epistemologist. Like her, Brunschvicg and Bachelard regarded the history of science as exhibiting different ways of organising data; as they put it, the mind has a history. Moreover, they did not see science as a monolithic activity. Against the positivistic aim of unification of science, they thought that philosophy could only reflect on one science at the time. Rather than a general philosophy of science, which, in Canguilhem's words, can only degenerate in 'futility and banality' (Canguilhem, 2015 [1965], p. 1098), they analysed particular sciences. Indeed, Bachelard put forward the concept of 'regional rationalism': for him different 'regions' of knowledge, these being different sciences, and also different areas of the same science, may exhibit different types of rationality (Bachelard, 1986 [1949], Chapter 4). They regarded current science as fragmented in terms of methods, and this could have appealed to Metzger, who studied the fragmentation of early-modern science. Nevertheless, Brunschvicg and Bachelard never thought that science, that is authentic knowledge, could be anything else but rational and secular. Imagination, emotions, desires and religious belief could simply not be part of science.

By contrast, Metzger found plenty of imagination and emotions in her sources. Should she then expel them from the history of science, as Bachelard suggested? She thought not, for two reasons. First, as mentioned in Chapter 4, she had established historical links between her sources, as practitioners read one another, accepted, modified or rejected others' ideas, and employed and modified one another's instruments. Second, she did not regard modern science as purely rational, nor did she think that we should make an effort to make it so. 'The most admirable inventions' she observed, may emerge in unlikely places, and imagination and emotions play an important role in science (Metzger, 1987 [1930], p. 121). She held, then, a much more nuanced and open conception of science even than the historical epistemologists, who regarded science as an open-ended process, subject to unforeseeable changes. This enabled her to have a far more charitable view of past theories and practices, and regard them as part of science and its history. She shared with Brunschvicg and Bachelard the view that one cannot predict where science would take knowledge. However, for her the development of science was even more inscrutable than for them, as not only rational beliefs, but all sort of concepts and practices could contribute to it.

There is, however, a more profound difference between Brunschvicg and Bachelard on the one hand and Metzger on the other. Their respective studies of science had different aims. The 'pedagogical' plan that Metzger praised in Bachelard's *Formation* aimed to show to the reader the pitfalls that the mind encounters in the study of nature. As he put it in *Psychoanalysis of fire* he intended to 'cure the mind of its happy illusions' (Bachelard, 1964 [1938], p. 4). For Brunschvicg and Bachelard, rationality is a value to be promoted. As a consequence, their histories of science are normative. For them, the norms of rationality and scientificity belong to current science. It is from the point of view of current science that they judged past science and constructed their histories, as I shall discuss more in detail in Chapter 7. Their philosophies depended on current science: as Bachelard put it, science 'creates philosophy' (Bachelard, 1984 [1934], p. 3), and philosophy must be reformed following the development of science (Bachelard, 1972 [1953], p. 135). Metzger was not a relativist, and certainly believed that current science was superior to past science, but at the same time she was not in the business of judging; rather, she aimed to understand. She was not interested in demarcating scientific thought from other types of thought. In fact, she was interested in the most diverse inquiries into nature; she valued all of them in different ways, and crucially thought that there was ample evidence that the practices and concepts that might look very unscientific to us had in fact given an important contribution to the development of scientific disciplines.

Just as she defended non-rational elements in scientific knowledge, so she believed that the historian needs to activate sympathy and intuition to penetrate past scholars' minds. In her milieu, her ideas were not seen favourably, and she received criticism from several quarters, not least from Émile Meyerson. We know of his criticism because of her repeated ripostes, both in her talks and in her own historical work. Indeed, she indicated Meyerson's 'De l'analyse des produits de la pensée' (Meyerson, 1934) as a condemnation of her method, in a tone that leads me to believe that he had their discussions in mind when writing at least some passages of his essay. There, Meyerson condemned 'introspection' and the '*a priori* method' in history, in favour of an objective analysis of the sources.[10] As the title of Meyerson's essay suggests, he advocated an examination of the 'products' of thought, that is written sources, rather than any attempt at reconstructing the mental states of past authors. Metzger replied that if she read her sources as 'objectively' as Meyerson suggested, she should discard them as meaningless. However, he, as mentioned above, had suggested that she use 'common sense'. The problem was that they had a different conception of the mind, and she thought that he assumed that the mind was fixed, a notion that she, along with Brunschvicg and Bachelard, did not accept. Their respective philosophies of mind were different and so were the intended objects of their histories: the 'products of thought' for Meyerson and the thought behind those products for Metzger.

Despite all these differences, Metzger, Meyerson, Brunschvicg and Bachelard studied the history of science in order to understand the mind. Metzger proposed the following philosophical goal for the history of science:

> 1) to understand the human mind better; and 2) by means of this very knowledge, to use our intelligence more wisely and less empirically than we have done thus far, building as we have done our scientific, philosophical and historical theories haphazardly. Or rather, without deceiving ourselves with grandiose projects or long-unfulfilled dreams, we shall like to manage to provide some services to scientists and philosophers.
>
> (Metzger, 2019 [1937])

The above quotation is a little ambiguous: should history of science be at the service of philosophers – and scientists – who would then extract lessons from it, or could history of science all alone go some way towards understanding the mind and showing the way forward? In other words, was the last sentence spoken out of modesty, or fully meant? Metzger was not completely clear on this point. The reality is that the distinction between history of science and philosophy was not clear at all precisely for scholars like herself who aimed at uniting history and philosophy. The study of the mind was the aim of a variety of projects in France at the time, notably of philosophers, including Meyerson, Brunschvicg, and Bachelard. As we shall see in the next chapter, Lévy-Bruhl also shared this aim with his fellow philosophers, but thought that it was ethnology that would give the answers.

## Notes

1 *The Monist* called it 'First International Congress for the Unity of Science' (Anonymous, 1935), but Metzger appears to have the official title: https://www.lib.uchicago.edu/e/scrc/findingaids/view.php?eadid=ICU.SPCL.USM, accessed 7/3/2018.

2  The talks in the Appendix of the present book were given by Metzger at the Centre de synthèse. Other talks include Metzger et al. (1930); Metzger (1932); Metzger (1934); these and the original French versions of those in the Appendix are reprinted in Metzger (1987). The discussions were published in *Archeion*, within the minutes of the meetings during which the discussions took place, generally separately from the talks, with the exception of Metzger et al. (1930), in which the discussion follows the talk (Metzger et al., 1933); (Metzger et al., 1935); (Metzger et al., 1936); (Metzger et al., 1937). Her audience included Mieli and Berr, and at different times other scholars, some of whom were frequent attendees. Among those who attended her talks we find Mieli's co-author Pierre Brunet (Mieli and Brunet, 1935), Abel Rey, the Swiss historian of medicine Henry Ernest Sigerist, Pierre Ducassé, the historian of science Bertha Bessmertny, 'J. Huizinga', whom I assume to be the famous Dutch historian Johan Huizinga, and on one occasion Lucien Febvre, who, although in positions of responsibility at the Centre, very seldom frequented the historians of science, with the exception of Abel Rey. Although there were women in the audience of her talks, they seldom joined in the discussion, with the notable exceptions of Metzger and Bessmertny. The latter also gave talks, for instance: Bessmertny-Heimann (1937).
3  Gad Freudenthal has more usefully compared Metzger's *a priori* with Gadamer's prejudice (Freudenthal, 1990).
4  Sarton, on the other hand, approved of Enriques' criticism of Duhem and overall thought highly of Enriques.
5  For instance, Enriques' insistence on the continuity of history of science, as well as on the unity of science, is a more traditional view than Metzger's complex picture (Enriques, 1934). The Italian edition of Enriques' book, which originally came out in 1936, has been republished with a number of critical essays, and includes Metzger's reviews in Italian translation (Enriques, 2004 [1934]).
6  See Chapter 4.
7  Metzger regards Bachelard's *La formation de l'esprit scientifique* above all as animated by a pedagogical aim, a view that I wholeheartedly share, see Chimisso (2001, 2008).
8  In this, Metzger certainly had an ally in Alexandre Koyré, who commented at one of her talks that 'we must make a great effort to understand how our far-removed predecessors worked; it seems to me that the human mind progresses in a manner that does not conform to the rules of common sense: here is history of science's first lesson' (Section d'histoire des sciences, 1935).
9  Heidelberger would probably partly disagree with my view. He has argued that 'Metzger's position is more akin to the anti-realist tradition which originated with positivism than to Meyersonian realism – Metzger's anti-positivist assertions notwithstanding' (Heidelberger, 1990, p. 151).
10  Incidentally, although Meyerson opposed positivism, his commendation of introspection recalls Auguste Comte's position (Bensaude-Vincent, 2008), and it is in line with Henri Berr's positivistic view of historical method.

# References

Anonymous. 1935. First International Congress for the Unity of Science. *The Monist*, 45, 308.
Bachelard, G. 1964 [1938]. *The psychoanalysis of fire*, Boston, Beacon Press.
Bachelard, G. 1968 [1940]. *The philosophy of no: A philosophy of the new scientific mind*, New York, Orion Press.

Bachelard, G. 1972 [1953]. *Le matérialisme rationnel*, Paris, Presses Universitaires de France.

Bachelard, G. 1984 [1934]. *The new scientific spirit*, Boston, Beacon Press.

Bachelard, G. 1986 [1949]. *Le rationalisme appliqué*, Paris, Presses Universitaires de France.

Bachelard, G. 1993 [1938]. *La formation de l'esprit scientifique: contribution à une psychanalyse de la connaissance objective*, Paris, Vrin.

Bachelard, G. 2002 [1938]. *The formation of the scientific mind*, Manchester, Clinamen Press.

Bensaude-Vincent, B. 2008. Meyerson critique ou héritier de Comte? *Dialogue-Canadian Philosophical Review*, 47, 3–23.

Berr, H. 1911. *La synthèse en histoire; essai critique et théorique*, Paris, Alcan.

Berr, H. & Febvre, L. 1932. History. *In*: Seligman, E. R. A. (ed.) *Encyclopaedia of the social sciences, vol. 7*, New York, MacMillan.

Bessmertny-Heimann, B. 1937. L'histoire des sciences dans l'Encyclopédie. *Archeion*, 19, 196–203.

Brunschvicg, L. 1912. *Les étapes de la philosophie mathématique*, Paris, Alcan.

Brunschvicg, L. 1922. *L'expérience humaine et la causalité physique*, Paris, Alcan.

Brunschvicg, L. 1934. *Les âges de l'intelligence*, Paris, Alcan.

Brunschvicg, L. 1936. History and philosophy. *In*: Klibansky, R. (ed.) *Philosophy and history. Essays presented to Ernst Cassirer*, Oxford, Clarendon.

Brunschvicg, L., et al. 1923. Histoire et Philosophie. Séance du 31 mai 1923. *Bulletin de la Société française de philosophie*, 23, 145–172.

Canguilhem, G. 2015 [1965]. *Philosophie et science. Résistance, philosophie biologique et histoire des sciences 1940–1965; Œuvres completes tome IV*, Paris, Vrin.

Chimisso, C. 2001, *Gaston Bachelard: Critic of science and the imagination*, London Routledge.

Chimisso, C. 2008. *Writing the history of the mind: Philosophy and science in France, 1900 to 1960s*, Aldershot, Ashgate.

Descartes, R. 1985 [1637]. *Discourse on method. The philosophical writings of Descartes*, translated by John Cottingham, Robert Stoothoff, Dugald Murdoch, Cambridge, Cambridge University Press.

Enriques, F. 1934. *Signification de l'histoire de la pensée scientifique*, Paris, Hermann.

Enriques, F. 2004 [1934]. *Il significato della storia del pensiero scientifico, a cura di Mario Castellana e Arcangelo Rossi*, Manduria, Barbieri.

Freudenthal, G. 1990. Épistémologie des sciences de la nature et herméneutique de l'histoire des sciences selon Hélène Metzger. *In*: Freudenthal, G. (ed.) *Études sur / Studies on Hélène Metzger*, Leiden, Brill.

Heidelberger, M. 1990. History of science and criticism of positivism: Émile Meyerson's and Hélène Metzger's views from a present-day perspective. *In*: Freudenthal, G. (ed.) *Études sur / Studies on Hélène Metzger*, Leiden, Brill.

Lalande, A. 1999 [1926]. *Vocabulaire technique et critique de la philosophie*, Paris, Presses Universitaires de France.

Metzger, H. 1918. *La genèse de la science des cristaux*, Paris, Alcan.

Metzger, H. 1926. *Les concepts scientifiques*, Paris, Alcan.

Metzger, H. 1932. Introduction à l'étude du rôle de Lavoisier dans l'histoire de la chimie. *Archeion*, 14, 31–50.

Metzger, H. 1934. La littérature scientifique française au XVIII$^e$ siècle. *Archeion*, 16, 1–17.

Metzger, H. 1938. *Attraction universelle et religion naturelle chez quelques commentateurs anglais de Newton*, Paris, Hermann.

Metzger, H. 1969 [1923]. *Les doctrines chimiques en France du début du XVIIᵉ à la fin du XVIIIᵉ siècle*, Paris, Blanchard.

Metzger, H. 1974 [1930]. *Newton, Stahl, Boerhaave et la doctrine chimique*, Paris, Blanchard.

Metzger, H. 1987. *La méthode philosophique en histoire des sciences. Textes 1914–1939, réunis par Gad Freudenthal*, Paris, Fayard.

Metzger, H. 1987 [1923]. Léon Brunschvicg, L'expérience humaine et la causalité physique. *In:* Metzger, H. 1987. *La méthode philosophique en histoire des sciences. Textes 1914–1939, réunis par Gad Freudenthal*, Paris, Fayard.

Metzger, H. 1987 [1929]. La philosophie de Émile Meyerson et l'histoire des sciences. *In:* Metzger, H. 1987. *La méthode philosophique en histoire des sciences. Textes 1914–1939, réunis par Gad Freudenthal*, Paris, Fayard.

Metzger, H. 1987 [1930]. La philosophie de Lévy-Bruhl et l'histoire des sciences. *In:* Metzger, H. 1987. *La méthode philosophique en histoire des sciences. Textes 1914–1939, réunis par Gad Freudenthal*, Paris, Fayard.

Metzger, H. 1987 [1934]. Federigo Enriques, signification de l'histoire de la pensée scientifique. *In:* Metzger, H. 1987. *La méthode philosophique en histoire des sciences. Textes 1914–1939, réunis par Gad Freudenthal*, Paris, Fayard.

Metzger, H. 1987 [1935]-a. Léon Brunschvicg, De la connaissance de soi. *In:* Metzger, H. 1987. *La méthode philosophique en histoire des sciences. Textes 1914–1939, réunis par Gad Freudenthal*, Paris, Fayard.

Metzger, H. 1987 [1935]-b. Réflexions sur l'École de Vienne. *In:* Metzger, H. 1987. *La méthode philosophique en histoire des sciences. Textes 1914–1939, réunis par Gad Freudenthal*, Paris, Fayard.

Metzger, H. 1987 [1936]. Émile Meyerson, *Essais. In:* Metzger, H. 1987. *La méthode philosophique en histoire des sciences. Textes 1914–1939, réunis par Gad Freudenthal*, Paris, Fayard.

Metzger, H. 1987 [1938]. Gaston Bachelard, La formation de l'esprit scientifique. Contribution à une psychanalyse de la connaissance objective. *In:* Metzger, H. 1987. *La méthode philosophique en histoire des sciences. Textes 1914–1939, réunis par Gad Freudenthal*, Paris, Fayard.

Metzger, H. 2019 [1935]. The tribunal of history and the theory of scientific knowledge (Appendix 2).

Metzger, H. 2019 [1936]. The *a priori* in scientific theory and the history of science (Appendix 3).

Metzger, H. 2019 [1937]. The philosophical method in the history of science (Appendix 4).

Metzger, H. et al. 1930. Les différents aspects de la même époque (lettres, sciences, arts) peuvent-ils être considérés comme autant de projections variées d'un même état d'esprit? Ou a contraire leurs modifications diverses ont-elles agi individuellement sur l'évolution de cette civilisation en général? Séance du 19 mars 1930. *Archeion*, 12, 375–379.

Metzger, H. et al. 1933. Centre international de synthèse, Section d'histoire des sciences, Séance du 18 janvier 1933 [Discussion of Metzger, 'Should historians ...']. *Archeion*, 15, 154–159.

Metzger, H. et al. 1935. Centre international de synthèse, Section d'histoire des sciences. Séance du 23 janvier 1935 [Discussion of Metzger, 'The tribunal of history']. *Archeion*, 17, 82–84.

Metzger, H. et al. 1936. Centre international de synthèse, Section d'histoire des sciences, Séance du 20 novembre 1935 [Discussion of Metzger, 'The A priori ...']. *Archeion*, 18, 75–79.

Metzger, H. et al. 1937. Centre international de synthèse, Section d'histoire des sciences, Séance du 21 avril 1937 [Discussion of Metzger, 'La méthode philosophique']. *Archeion*, 19, 254–257.

Meyerson, É. 1930 [1908]. *Identity and reality*, London, New York, Allen & Unwind, MacMillan.

Meyerson, É. 1931. *Du cheminement de la pensée*, Paris, Alcan.

Meyerson, É. 1934. De l'analyse des produits de la pensée. *Revue philosophique*, 118, 135–170.

Meyerson, É. 1936. *Essais*, Paris, Vrin.

Mieli, A. & Brunet, P. 1935. *Histoire des sciences. Antiquité*, Paris, Payot.

Section d'histoire des Sciences. 1935. Séance du 23 Janvier 1935. *Archeion*, 17, 81–84.

# 6 History of science, ethnology and religion

> Lévy-Bruhl has always encouraged me in my work, and ... has always agreed
> to be the first reader of my work
>
> (Metzger, Letter to Sarton, 14.4.1927)

## The past is a foreign country

Hélène Metzger had a very direct connection with the founder of the discipline of ethnology in France, namely Lucien Lévy-Bruhl, who was her uncle. His interest in her work is testified by Metzger herself as exemplified in the quotation above; we also have a letter in which he requested conference registration forms on her behalf (Lévy-Bruhl's Letter of 21.12.1921, in (Merllié, 1993, Appendix)). She did not mind compiling the indexes of his monographs, for which Lévy-Bruhl acknowledged her in his books (Lévy-Bruhl, 1996 [1927], 'Avantpropos'); (Lévy-Bruhl, 1963 [1931], p. xiii), whereas she greatly objected to Meyerson's requests for the same services (Freudenthal, 1990, Metzger to Sarton, Letter of 14.4.1927, p. 255). Incidentally, Lévy-Bruhl's political leanings were in tune with Metzger's: he was a close friend of the socialist leader and fellow Dreyfusard Jean Jaurès, of whom he wrote a biography (Lévy-Bruhl, 1924a); he presented his own political views in 'L'idéal républicain' (Lévy-Bruhl, 1924b). These fortunate circumstances, however, only facilitated Metzger's knowledge of the emerging *ethnologie*, which she would have surely acquired anyway. No historian of science would have overlooked Lévy-Bruhl's theory of mentality; nor would his fellow philosophers at the Sorbonne. Lévy-Bruhl retired as professor of history of philosophy, and a great number of his publications are in this discipline, including those on Comte, Hume, and German and French philosophy (Lévy-Bruhl, 1890); (Lévy-Bruhl, 1894); (Lévy-Bruhl, 1899b); (Lévy-Bruhl, 1899a); (Lévy-Bruhl, 1903 [1900]); he also published on ethics (Lévy-Bruhl, 1884); (Lévy-Bruhl, 1903); (Lévy-Bruhl, 1905 [1903]). His late works on mentality had a resonance that is perhaps difficult to imagine nowadays, when they have been all but forgotten, or long rejected, although with exceptions, one of which will be briefly discussed in Chapter 8.[1]

His fellow philosophers who studied the mind were not alone in discussing his work.[2] As Metzger attested:

> [Lévy-Bruhl's theory] has immediately gone beyond the small circle of the specialists which it seemed to address: it has been read, discussed, commented on, employed, not only by ethnologists, travellers and colonial administrators; but also by philosophers, psychiatrists, psychologists, pedagogues, sociologists and historians.
>
> (Metzger, 1987 [1930], p. 117)

Lévy-Bruhl was already in his fifties when he decided to switch his interest from European philosophy to African, Polynesian and Australian civilisations. His description of his first encounter with non-European thought, which sparked his interest in different ways of thinking, is rather similar to Metzger's description of her own encounter with her early-modern sources. He narrated that he received a French translation of a Chinese historian's work, but no matter how much he studied it, he could not fully understand it. His further efforts with other Chinese books led him to believe that the authors reasoned in a different way from him. Similarly, Metzger was baffled by past writings of European scholars. She experienced the same puzzlement as Lévy-Bruhl did, the only difference being that her authors were chronologically, rather than geographically, distant. His experience with that Chinese book led him to investigate different mentalities, focussing on what at the time were called, and believed to be such by many, 'primitive' societies in Africa, Australia and Asia (Lévy-Bruhl, 1910); (Lévy-Bruhl, 1925 [1922]). He did so by analysing ethnographic reports, which for him could not be taken at face value: it was necessary to understand the minds that had produced those claims (allegedly) transcribed in those reports. Similarly, Metzger aimed to understand how the minds of people for whom her sources were easily intelligible worked, so that she too could understand them.

Metzger believed that Lévy-Bruhl's work was very relevant to the history of science, and indeed envisaged a collaboration between ethnology and history of science which would afford 'a more precise view of the structure of the human mind' (Metzger, 1987 [1930], p. 123). There was a clear obstacle, however. Lévy-Bruhl, even if not decisively, objected to the unity of the human mind, especially in *La mentalité primitive* (Lévy-Bruhl, 1925 [1922]), in an apparent break even with the Durkheimians, including Marcel Mauss, the co-founder, with Lévy-Bruhl himself, of the Institut d'ethnologie.[3] For Metzger it was not a matter of differences between Europeans and peoples who, in her words, 'our pride of civilised people qualifies, without doubt erroneously, as savages' (Metzger, 1987 [1930], p. 115). It was rather a matter of differences between past and present ways of thinking between which her own narratives created links and a shared history. How could then Lévy-Bruhl's theory be of any service to her? Metzger was an attentive reader of his work, and must have had many chances to have discussions with him. This enabled her on the one hand to give a precise presentation of Lévy-Bruhl's theses, and on the other to extract from his work points of reflection and concepts that she could use. In her presentation, Lévy-Bruhl had never really been wedded to the hypothesis of the disunity of the human mind, and indeed he had dedicated many pages of his *Les fonctionnes mentales dans les sociétés inférieures* (Lévy-Bruhl, 1910) (translated as *How Natives think* (Lévy-Bruhl, 2015 [1910])) to the presence of 'prelogical participation' in 'current mentality', by which she must have meant modern European mentality (Metzger, 1987 [1930], p. 118). She probably also had access to Lévy-Bruhl's evolving views, which in 1938 brought him to write in his *Carnets* (*Notebooks*) that the logical structure of the mind is the same in all known human societies (Lévy-Bruhl, 1975 [1947], p. 49)[4]. A few words must be spent on the terms 'prelogical' and 'participation'. Lévy-Bruhl suggested that the people whom he called primitives had a prelogical mind, by which, he said, he did not mean either 'antilogical' – as that would be unintelligible – or 'alogic', as it would only be confusion. Rather, he meant 'indifferent, in certain cases, to contradiction, because it is mystical, and it obeys both to the law of participation and to the principle of (non)contradiction' (Lévy-Bruhl et al., 1923, pp. 631–632). The 'law of participation' for Lévy-Bruhl is a 'general law, a common foundation for those mystic relations which primitive mentality so frequently senses in beings and objects …'. All those relations would 'involve a "participation" between persons or objects which form part of a collective representation' (Lévy-Bruhl, 2015 [1910], loc. 910). His examples included that of the Trumai, a tribe of Northern Brazil, who reportedly claimed to be both humans and aquatic animals. This and many other examples led Lévy-Bruhl to think that that style of thinking disregarded the 'opposition between the one and the many, the same and another, and so forth' (Lévy-Bruhl, 2015 [1910], loc. 924).

Metzger was very interested in the mentalities that Lévy-Bruhl analysed. This was because, far from considering them exotic ways of faraway lands, she was reminded of types of reasoning that she had encountered in her sources, notably in Renaissance medicine and alchemy. She was not alone in connecting Lévy-Bruhl's concept of primitive mentality with past ways of thinking. In fact, that was the main way in which philosophers and historians employed it. Just to mention a few examples among many: Brunschvicg employed Lévy-Bruhl's concept of prelogical mentality to give his account of the early concept of number (Brunschvicg, 1912, Chapter 1); Abel Rey did the same in his entry in the *Encyclopédie française* on the evolution from primitive to current thought (Rey, 1937); Gaston Bachelard compared Francis Bacon's peculiar medical remedies with those of Melanesian people as described by Lévy-Bruhl; for him the latter's concept of participation would explain both (Bachelard, 2002 [1938], pp. 150–151); (Lévy-Bruhl, 1925 [1922], pp. 384ff). The philosophers and historians who employed

Lévy-Bruhl's concept of mentality in order to describe the past of science translated his synchronic model into a diachronic model, by comparing the mentalities behind natural philosophy, or the first concepts of number, with those of twentieth-century Melanesians and Africans. Their application posed a philosophical problem: were the latter's mentalities similar to those exhibited by past Europeans because they were at an early stage of a development that follows the same path regardless of place or culture? Or are different mentalities independent of one another? This dilemma was in fact the fundamental disagreement between Lévy-Bruhl and the Durkheimians, to whom he was nevertheless close. Durkheim himself described this particular disagreement in his review of Lévy-Bruhl's *How Natives Think*, in which he compared it with his own work *The elementary forms of religious life* (Durkheim, 1909–12). He emphasised that he and Lévy-Bruhl shared the same fundamental research principles, and both thought that different mentalities have emerged in history (Durkheim, 1909–12, p. 35). However, Lévy-Bruhl opposed religious and primitive mentality to modern and scientific mentality as two different and autonomous mentalities. By contrast, Durkheim held that 'the highest and more recent forms are not opposed to the more primitive and inferior forms, rather [the former] are born of the latter' (Durkheim, 1909–12, p. 37). Lévy-Bruhl's and Durkheim's distinct models of the relationship between mentalities, and their connection with historical development, were often received together, adapted and modified by historians, philosophers, psychologists and other intellectuals. Lévy-Bruhl's synchronic view was at odds with mainstream philosophy, especially with Comte's law of three stages, according to which civilisations start off with the theological stage, go through the intermediary metaphysical stage, and finally reach the positive stage. Lévy-Bruhl did not regard indigenous people's ways of thinking as a previous stage in a historical development that would lead to the present mentality of modern European people. There was, however, some confusion in the reception of his ideas. This was not helped by his use of the term 'primitive', which suggests beginnings. On the other hand, he often stressed that this term was not appropriate, and proposed to change attitude towards indigenous populations' way of explaining phenomena in the following way:

> let us abandon the attempt to refer their mental activity to an inferior variety of our own. Rather let us consider these connections [among phenomena] in themselves, and see whether they do not depend upon a general law, a common foundation for those mystic relations which primitive mentality so frequently senses in beings and objects ... I shall call [this] principle ... the law of participation.
>
> (Lévy-Bruhl, 2015 [1910], loc. 906)

Lévy-Bruhl's synchronic model supported the project that philosophers like Brunschvicg and Bachelard pursued, and their discontinuous view of history of science. In fact, Brunschvicg chose Lévy-Bruhl's model precisely because, according to him, it rejected the Durkheimians' assumptions that the mind and its categories are the same in all times and places; and that different ways of thinking are just stages of a universal mental evolution (Brunschvicg, 1932). For Brunschvicg, current scientific rationality is the highest, though provisional, point of human thought. Precisely because there is no guarantee of intellectual progress, for him scientific rationality should be preserved and promoted against superstition and religion. Bachelard employed the synchronous model in a clear-cut manner, as he argued that primitive and scientific mentalities co-exist not only at the same time, but in the same civilisation, indeed in the same individual. Primitive or pre-scientific mentality for him is the poetic, imaginative and affective approach to reality that a healthy human being would cultivate alongside a scientific, discursive and rational mentality. These two ways of thinking, however, must be kept separate; in fact, the latter develops in opposition to the former (Bachelard, 2002 [1938]); (Bachelard, 1964 [1938]). Against this background, Metzger's reception of Lévy-Bruhl appears to be original and nuanced. While she employed his theory of primitive mentality, and his law of participation in particular, she considerably revised them, in a different manner from Brunschvicg, Bachelard, and indeed most of the philosophers in her milieu.

## Ethnology and science

Metzger was confident that Lévy-Bruhl's theories, suitably interpreted, could 'help the historian of science penetrate the mind of scientists whose work he studies', despite the differences in their respective subject matter (Metzger, 1987 [1930], p. 123). Both she and Lévy-Bruhl aimed to understand minds that at first look alien, and to isolate the concepts and assumptions that govern them. In particular, Metzger based her central concept of active analogy, analysed in Chapter 1, on Lévy-Bruhl's law of participation (Metzger, 1926, p. 35); (Metzger, 1987 [1930], p. 123n); (Metzger, 1927); (Metzger, 2019 [1936]). In her presentation of her concept of active analogy, Metzger quoted a passage from *How Natives Think* in which Lévy-Bruhl wrote that the objects in a relationship of participation 'give forth and receive mystic powers, virtues, qualities, influences, which make themselves felt outside, without ceasing to remain where they are' (Lévy-Bruhl, 2015 [1910], loc. 920). Lévy-Bruhl cited various examples, reported by Metzger, including that of the newborn child who would feel the effects of what its father does, and that of the man whose fortunes in battle would depend on the behaviour of his wife back at home (Lévy-Bruhl, 2015 [1910], loc. 935); (Metzger, 1926, p. 37); (Metzger, 2019 [1936]). Other examples given by Lévy-Bruhl involve relationship between objects, and between people and objects or events. He mentioned the links that 'native' people establish between abundance of game, fish and fruit and the performance of particular ceremonies, or the well-being of a sacred personality (Lévy-Bruhl, 2015 [1910], loc. 937). If the beliefs in this type of interactions only belonged to cultures that stood in opposition to European ways of thinking, then Lévy-Bruhl's work could not have been relevant for scholars who, like Metzger, studied European intellectual history. She, however, cited medieval magical practices, in which a relationship of sympathy is created between a person and a wax doll representing them, so that the person would suffer what is done to the doll (Metzger, 1926, p. 38), in a manner that in fact recalls voodoo practices. These practices to her appeared to be based on the concept of active analogy, in a form particularly close to Lévy-Bruhl's law of participation. She detected other similarities between the mentalities described by Lévy-Bruhl as primitive and those of the authors of her sources. For example, Lévy-Bruhl wrote that the indigenous people of Australia often covered their bodies with red ochre because the latter is like blood, indeed for them it is blood, and therefore confers strength. This practice for Metzger resembled that of sixteenth-century scholars for whom 'everything that was red would heal a bleed and thus save the body' (Metzger, 2019 [1936]). Active analogy for her was at work in both cases. Her comparisons led her to conclude that there was no evidence in support of the opposition between European and 'native' ways of thinking; she wrote that there is no 'abyss' between them (Metzger, 1987 [1930], pp. 118–119). She disagreed with Lévy-Bruhl's conclusion, but not with his method: he thought that the unity of the human mind should not be an *a priori* assumption but rather the result of empirical research. She approved of this, as shown by her criticism of the philosophers who read sources superficially, and used them as justification of ideas that they had already assumed at the beginning.[5] An attentive study of sources, and comparisons with Lévy-Bruhl's descriptions of 'native' thought, had convinced her that both ethnology and history of science lead to the confirmation of 'the unity of the mental structure of the human species' (Metzger, 1987 [1930], p. 123).

Metzger's examples of medieval sorcery, and of astrology (Metzger, 1926, p. 49), might suggest that she saw a connection between current African and Melanesian mystical thought and past European thought. This was a widespread view, and a view that was consistent with the positivistic model of stages of civilisation. However, Metzger detected the concept of active analogy, and indeed of the other types of analogy, not only in sorcery and astrology, but in Stahl's chemistry of affinities. Was the mystical way then an enduring framework that involved all civilisations, but that was progressively eliminated by science, as her teacher Brunschvicg suggested? Her respect for Stahl would already suggest that she did not accept that view, but her other examples make it clear: she mentioned theories behind vaccines and endocrinology (Metzger, 1987 [1930], pp. 123, 125). She thought that, although the law of participation as an all-encompassing way of interpreting nature might have disappeared from modern thinking, still many aspects of participation survive in science. Her favourite example was Newton's law of universal gravitation, which had re-established action at a distance after Cartesian rationalism had banned it, as I have mentioned. Metzger thought that Lévy-Bruhl had 'erroneously' labelled the way of thinking that he described as 'primitive mentality'. She judged many of his analyses to be accurate, yet what she saw in them was not primitivity, but rather spontaneity. The use of active analogy for her belongs to 'spontaneous thought'. When thinking spontaneously, people, including scientists, are not very concerned with coherence, or the risk of contradiction. Spontaneous thought is not fully logical and certainly not systematic; it follows new ideas along diverse, and even contradictory, paths, and is open to many possibilities (Metzger, 1987 [1930], p. 120). When reading sources that proceed from spontaneous thought, the historian should not attempt to determine their logical structure independently of their contents. She proposed a more holistic reading, so to speak, which includes content and context, and does not overlook the emotions involved in the production of the text in question (Metzger, 1987 [1930], p. 122). It goes without saying that, alongside spontaneous thought, there is the logical and methodical way of thinking that would check and revise the intuitions of spontaneous thought. This for her had been the great achievement of Newton, namely, to have organised into a rigorous mathematical framework the intuition of action at a distance, which indeed comes spontaneously to our minds. In contrast with philosophers like Brunschvicg and Bachelard, she argued that the elimination of spontaneous thought would in fact hinder science. For her, without it, there would have been neither the theory of universal gravitation, nor the theory of evolution, nor vaccines. It seemed to her that biology and medical sciences particularly benefit from a generous dose of spontaneous thought (Metzger, 1987 [1930], pp. 122–123). Interestingly, she employed an organic image to show her view of the relationship between spontaneous thought and logic:

> Clear, limpid and logical thought ... would not produce its most beautiful fruits on its own plant. It needs to be grafted on a wilding in order to become truly productive.
>
> (Metzger, 1987 [1930], p. 121)[6]

In short, Metzger regarded so-called primitive thought as spontaneous thought that is found everywhere: among peoples of Africa and Asia, as well as Europeans; in any time: in early-modern natural philosophy as well as modern science; and in any activity: in sorcery as well as in science.

There are, however, several further differences between her view and Lévy-Bruhl's. First of all, for her it is not the case that all spontaneous thought is mystical, although the latter without doubt is a form of spontaneous thought. She held that spontaneous thought is a much larger category than mystical thought, as it is all thought in its free and creative phase, not fully disciplined by reflection and strict logic. Moreover, for Metzger, active analogy is not dominant in all instances of spontaneous thought in the way the law of participation is in Lévy-Bruhl's primitive thought. As we have seen when discussing *Les concepts scientifiques*, concepts of evolution and substance among others for her have played crucial roles in different periods and domains of the study of nature. In any case, in her view, human thought employs a number of concepts at the same time; at any one time, some of these concepts dominate, at other times these same concepts would be rejected, and then sometimes be employed again within a novel context. It is in light of her view of spontaneous thought and the role of concepts in human understanding that we should interpret her remark that the difference between primitive and 'civilised' thought was a matter of 'degree' (Metzger, 1987 [1930], p. 119). It is not a matter of degree in the sense of a step in the path of progressive civilisation, but rather of degree of use of spontaneous thought, and of certain concepts rather than others.

Lévy-Bruhl declared the transmission of powers and influences of primitive thought 'incomprehensible to us' (Lévy-Bruhl, 2015 [1910], p. 920). It could be argued that he was not speaking literally, as he dedicated thousands of pages precisely to the explanation of how primitive thought worked. Nevertheless, he never proposed that he could go 'native' and think as a member of the societies he read about. This is in contrast with Metzger's aim to become a contemporary of the scholars whose work she studied. Her position was different from his for two reasons. The first is that, as discussed, she did not think that the concepts used by past scholars and so-called primitives were unknown to modern people. However great the gap between herself and past scholars, it could be bridged precisely because no matter how differently people from other times and places saw nature and organised phenomena, they still used categories, inferences and emotions that we can recognise. The second reason is that she allowed for spontaneous thought to be part of scholarly inquiry, including science and history. In so doing, she had an additional method and indeed another type of communality with minds from other cultures. As we know, her method of active sympathy was not shared by the scholars in her milieu. It was not shared by Meyerson, who as we know, all but mocked it by calling it introspection. Nor was it shared by Lévy-Bruhl, who aimed for a positive and fully rational approach to his sources. Yet, Lévy-Bruhl's work was crucial for her. Although she believed in the unity of the mind, for her the varieties of ways in which human beings employ concepts is so diverse that great scholarly effort, which included all the historian's faculties, is needed just to make some progress in the understanding of past minds. Lévy-Bruhl analysed types of reasoning, notably based on analogy, that historians of science often neglected, but that she thought were critical to understand past sources.

## Religion and science

For many philosophers close to Metzger, scientific progress went hand-in-hand with the advance of secularisation. This is a clear theme in Brunschvicg's philosophical histories of mathematics and causality (Brunschvicg, 1912,

Brunschvicg, 1922). In various manners, his view was shared by other philosophers, including Rey and Bachelard, by Lévy-Bruhl, by sociologists; and by the historians of the Centre de synthèse. History of science as an academic discipline in France was largely the creation of secular intellectuals. There were examples of historians and philosophers of science who regarded religion and science as compatible and indeed part of the same historical development, notably the Catholic and conservative Duhem, who however died in 1916, before the creation of academic courses and degrees in history of science. Intellectually and socially, as an apparently secular Jew and 'almost a socialist', Metzger was very much in tune with the dominant views in the institutions where she worked, namely the Centre de synthèse and the Institut d'histoire des sciences at the Sorbonne. Yet, her view of the relationship between religion and science differed from that of most of her colleagues. She emphasised the role that religion in all its forms has played in the history of science including mysticism and theology, the latter understood both as a system of beliefs to which chemists consciously subscribed, and as a collection of assumptions that they accepted without discussions. Paracelsus' mystical doctrine of signatures, originally employed by medics, was also embraced by chemists, who extended its validity to all substances (Metzger, 1969 [1923], pp. 156–157). Van Helmont rejected Aristotle and Galen's cosmology because it was pagan. For him, he was simply blasphemous to suggest that God is just an 'unmoved mover' that acts purely mechanically on matter (Metzger, 1969 [1923], pp. 166–168). Metzger concluded that in van Helmont's work, the alliance of religion and science was aimed at the destruction of Aristotle and Galen's philosophy, which still attracted so many scholars despite its pagan origin (Metzger, 1969 [1923], p. 183). No doubt, many of Metzger's colleagues considered van Helmont above all a mystic. However, she argued that there was strong continuity between the Flemish scholar's 'spiritualist' chemistry and Robert Boyle's corpuscularism. For her, not only did van Helmont's critical approach prepare the way for Boyle's scepticism, but both scholars also aimed to find the fundamental sameness of the apparently most diverse substances (Metzger, 1969 [1923], p. 186). Metzger quoted Boyle positively referring to van Helmont, in order to show the historical link between the two (Metzger, 1969 [1923], p. 187). For her, genuine experimental chemistry was born under the indirect influence of Renaissance 'mystic and theosophical doctrines' (Metzger, 1969 [1923], p. 193).[7] She also discussed specific changes in chemical theories due to religious reasons. An example is the rejection of the transmutation of metals, which had been the core of the alchemists' research programme. The latter was based on the idea of the perfection of gold, and the relative and hierarchical imperfection of the other metals. At the time when Cartesianism was dominant, scholars, including those who did not accept Descartes's overall philosophy, rejected that an extended substance could be more perfect than another. God created metals 'to stay what they are', just as the rest of creation. Change is only due to movement, and God is the first cause of movement (Metzger, 1969 [1923], pp. 138–139).

On the other hand, the same Cartesian metaphysics condemned the alkali-acid dualism as too complicated, and as such not consistent with God's perfection (Metzger, 1969 [1923], p. 214). Metzger also analysed the controversy between Nicolaas Hartsoeker and Leibniz on the indivisibility of atoms, and emphasised the centrality of God's will in their respective arguments (Metzger, 1969 [1923], p. 440). The Bible itself was an important source for many chemists across centuries. Metzger pointed out that the seventeenth-century scholar Johann Joachim Becher, a reference for Stahl, argued that minerals must have come from human beings, as they are not mentioned in the Bible (Metzger, 1991 [1930], p. 38). The examples could be endless, but arguably hardly useful, for, as Metzger herself put it, 'for long periods', it was impossible to separate chemical theory from metaphysics, biology, medicine, and crucially here, theology (Metzger, 1974 [1930], p. 8). Nobody would have argued with her judgement, but there would have been no agreement in her milieu about the period in which the separation between the sciences, on the one hand, and theology and religious beliefs on the other occurred. More importantly, did she think that religion was an obstacle to scientific progress, an irrational residue to be eliminated? I shall look for an answer to this question in two different ways. The first is to consider the importance that, for her, religion had in Newton's system, and indeed the importance of Newton's system for religion. I shall then take a more general look at Metzger's view of the role of religion in science.

I have already mentioned that Metzger regarded the law of universal gravitation as a 'return' to the concept of active analogy within a mechanistic view of nature. Did active analogy also carry religious meanings? Metzger thought so and set out to demonstrate it. Although originally God was at the core of the Cartesian worldview, Metzger pointed out that it could also be interpreted as a version of materialism, and indeed so it was in Britain in Newton's time (Metzger, 1938, pp. 3–4). Newton introduced a force into the universe that, against mechanistic philosophy, depends on bodies' own substance and is exerted where bodies are not. The mutual attraction of celestial bodies was seen as proceeding from God's will (Metzger, 1938, pp. 4–5). She wrote:

> The law of universal attraction seemed to make directly perceptible to our mind the permanent action of an omnipotent, omnipresent and perpetually acting God, who preserves with infinite wisdom the harmony of a world which he has admirably created.
>
> (Metzger, 1938, p. 6)

Metzger was keen to show that Newton himself saw his law in this light. She quoted, it must be said rather loosely, his letters to Bentley, in which Newton wrote that he was extremely glad that the principles of his system could help men to believe in a god, and that his system required a 'supernatural power' (Metzger, 1938, p. 60).[8] Newton's God, she pointed out, is a personal God, who does no reveal himself to us through our intellect, but rather directly through our emotions (Metzger, 1938, p. 66). In this case, we do not see theology as doctrine at work, but rather religious experience, in which emotions play an important role. Once again, Metzger emphasised that science is not created only by reason, and that its progress does not exclusively depend on it.[9]

Metzger's subject matter in *Attraction universelle et religion naturelle chez quelques commentateurs anglais de Newton* was judged fresh and original in her time. Her reviewer for *Isis*, the MIT historian of science Giorgio de Santillana, wrote:

> The Newtonian era is a crucial one in the history of civilization; it has the good fortune of attracting at present a number of competent investigators. We need only mention the names of Miss Marjorie Nicolson, Ch. Monroe Coffin, Robert K. Merton. But all of these authors are concerned mainly with its precedents. It is no less essential to see what was the influence of Newton's discoveries in his day. Modern scientists are apt to say that they just stood for a new 'positive' approach. Actually, like most important ideas, they acted through their intimations, ambiguities, and overtones. Indeed, we may say with the author [Metzger] that the force of Newton's ideas lay in his 'inspiration'.
>
> (Santillana, 1940, 145)

De Santillana stressed Metzger's complex view of science, which, he wrote, does not exclude, indeed emphasises, the tortuous and not always rational path that science takes. Considering her milieu, it was even more remarkable that she had decided to show the impact of science on religion. Certainly there was a practical reason for it: the book came out of her lectures at the École pratique des hautes études, in the department of religious sciences. This circumstance, however, does not detract from the interest of her choice, which is consistent with her overall view of science. She thought that she had isolated a particular historical moment, in which a scientific discovery, namely universal attraction, was greeted with 'immense religious enthusiasm'. Only half a century later, she added, religious thought had become rather indifferent to fundamental scientific discoveries (Metzger, 1938, p. 200). Metzger wrote that Richard Bentley (1662–1742) regarded gravitation as the result of God's immediate will (Metzger, 1938, p. 82). Both Richard Bentley and William Whiston (1667–1752), she claimed, regarded Newtonianism as a solution to the impasse created by Cartesian philosophy and its purely mechanical view of the universe. The latter was seen, as Metzger put it paraphrasing Bentley, as an 'atheistic hypothesis' (Metzger, 1938, p. 87).

I shall not dwell on her detailed discussion of the individual commentators on Newton. However, it is worth noting her insistence on the limited role that rationality played in the religion of those commentators on Newton.

For instance, in the context of the dispute between the Newtonian Samuel Clarke and Leibniz, she emphasised that their contemporaries regarded Newton and Leibniz's gods as profoundly different. Newton's God is 'based on irrationality, contingency, finalism' and is talked about as one would of an artist. By contrast, Leibniz's God, just like Descartes', is the guarantor of universal reason (Metzger, 1938, p. 125). Doubtless, she was aware that she was not the first to point out the role of religion in Newton's thought; she relied on a number of sources, notably Edwin Burtt's *The metaphysical foundations of modern science* (Burtt, 2003 [1924]). She was not even the first in her milieu to discuss the opposition of the Cartesians and Leibnizians' God on the one hand, and the Newtonians' God on the other. Indeed, in a brief footnote, she cited Léon Brunschvicg saying that Descartes's God is the God of reason, whereas Newton's God is a man who has grown as large as the universe (Metzger, 1938, p. 76). However, she did not share his negative evaluation of the Newtonian God and of his role in science. For Brunschvicg, Newton's theology was just an expression of everyday anthropomorphism (Brunschvicg, 1922, pp. 246, 584). He recalled a long tradition of criticism: the 'French Cartesians' Jean Bernoulli and Fontenelle had judged the idea of attraction 'unintelligible'; and Comte had relied on their view to condemn Newton's philosophy as 'primitive' (Brunschvicg, 1922, p. 336). For him, Kant managed to rescue, so to speak, Newtonianism from anthropomorphism and religion. Newton had placed space and time on a 'transcendent, that is to say imaginary' plane, whereas Kant conceived them as human in origin (Brunschvicg, 1934, p. 102). In order words, with Kant God had been eliminated from science. Arguably, Metzger accepted science more fully even than philosophers like Brunschvicg and Bachelard. They thought that philosophy should follow science; however, they assessed theories and worldviews according to norms of rationality and secularism. By contrast, Metzger did not judge whether a theory or a worldview was sufficiently rational or secular; rather, she observed that religion, the imagination and spontaneous thought have played a role in the development of science. Her non-normative approach to history of science led her to evaluate the relationship between science and religion, and science and spontaneous thought, in a way that set her apart from most of her colleagues.

## Science, religion and totalitarianism

In the work she published in her lifetime, Metzger only discussed religion as a historian, in the same way in which she described any other system of beliefs and worldview. She did not express any views on the roles that science and religion should play in current society. She hardly mentioned religion in her letters, and I think we can assume that she was secular, as were many of the people around her. In this context, her last manuscript, and in particular its title, *La science, l'appel de la religion et la volonté humaine*, immediately attracts one's attention. This is a draft outline of a planned philosophical book about science, religion and human will. As mentioned in the Introduction, she worked on it in 1941–42, while she was in Lyon, after fleeing Nazi-occupied Paris. The war, her deportation and murder prevented her from finishing it. Her brother, Adrien Bruhl, who curated its posthumous publication, presented her work as the 'preamble' to a study on the problem of God in current thought, in turn part of a planned book on Jewish monotheism (Bruhl, Adrien in (Metzger, 1954, p. 5)). Her manuscript appears to be an extended plan of a monograph, consisting of ten chapters. She herself described the topic of her book as the relationships between science and religion (Metzger, 1954, p. 8). The reader might then be surprised by the relatively small role that religion plays in it. Indeed, the scope of the book appears to be much broader. The situation in which it was written was clearly relevant, although never mentioned in any detail. She wrote that the 'public' was becoming increasingly more gullible, and accepted or fought ideologies, superstitions and truths, without any tool to distinguish truth from error (Metzger, 1954, pp. 39ff). This state of confusion for her had been fertile ground for the exaltation of instincts and irrational passions, which had been accepted as the basis for action. People had given up the guide of their conscience ('the inner *gendarme*') to submit willingly to a cruel and pitiless external '*gendarme*', regarded as the only one able to impart order on society and minds (Metzger, 1954, pp. 39–40). Rather than investigating past mentalities as she had done in her previous work, here she set out to analyse mainstream ways of thinking of her time. She aimed to isolate the roots of the current situation, and, as a historian of science, she detected four of them: the image of the scientific method held by philosophers, scientists and indeed the general public; biological and sociological explanations of human conscience; scientific and philosophical determinism; and the ideology of industrial progress and the exaltation of human power and instincts (Metzger, 1954, pp. 8 and passim). She was aware that her criticism could suggest that she was attacking science as a whole; she was therefore careful to dispel this potential misunderstanding. She wrote that by no means she condemned 'scientific and industrial activity', nor did she follow Rousseau in asserting that the thinking man is a depraved man, or Ferdinand Brunetière in declaring science a failure.[10] In fact, she wrote that she demanded 'more knowledge, more critique, more science' in order to combat philosophical and religious ideas that place 'the sources of our aspirations' only in a 'vague and superficial sentiment', (Metzger, 1954, pp. 40–41).

She had dedicated much of her intellectual effort to showing the varied ways in which scholars studied nature and made discoveries. The notion that there is an infallible and single scientific method that fully unlocks the secrets of nature is for her simply a misrepresentation of scientific practice that fuels the 'naïve aspiration' for complete knowledge (Metzger, 1954, p. 9). Once again, she fought the positivistic view of the mind as a 'passive receptacle' of an objective state of affairs (Metzger, 1954, p. 9, 31).

Although she labelled as 'positivist' the view of scientific method that she criticised, her target arguably was broader than that. She attacked the representation of the ideal scientist as devoid of desires, will and passions in his 'search for the truth', imperturbable and indifferent to his results (Metzger, 1954, p. 43–44). She was indeed inclined to regard what she saw as positivism and objectivism as just 'dialectical stages' which would be overcome by new ideas. For her, people cannot acquire knowledge if they have no passion to investigate, or desire to know.[11] It was not only a matter of method or attitude, however, but also of content of two sciences in particular: biology and sociology. For her, biologists, especially following the success of the theory of evolution, claimed to explain all human existence in terms of heredity, adaptation, natural selection and struggle for existence (Metzger, 1954, p. 18). She wrote that there was no space left for finality, and for a 'destiny superior to animal life' (Metzger, 1954, p. 17). Biology presented life, rather than as a struggle for wisdom, justice and happiness, as a struggle based on egoism, in which the best adapted creatures win (Metzger, 1954, p. 48). Similarly, sociologists appeared to her to explain away human inclinations as well as social change in terms of scientific laws. Her response, which she never developed in full, was a sort of moral rebellion. She wrote that society tends to conservatism, against which the underprivileged and people driven by a sense of justice fight. Sociology for her was destroying those fighters' hopes. In other words, she strongly objected to the negation of human freedom that sociology and biology appeared to suggest (Metzger, 1954, pp. 48–49).

Consistently with her criticism of the current trends of those sciences, she planned to dedicate a chapter to the critique of determinism. She recalled Laplace, who claimed that the knowledge of the immutable laws of nature and the state of the universe at a given moment would enable us to know all that preceded that moment and, crucially, all that would follow it. Her planned response echoed the central thesis of Emile Boutroux's book (Boutroux, 1916 [1874]), namely that natural laws are in fact contingent (Metzger, 1954, pp. 51–52). For her, the biology and sociology of her time were aimed at presenting nature and society as they objectively are, were and will be. Her view of knowledge contradicted their aspiration to complete objectivity. Like Brunschvicg, she regarded knowledge as mutual transformation of mind and object. Consistently, she opposed the epistemologies that she called 'contemplative'. In her analysis, they assume the fixity of the mind and of the object, and reduce science to a mere record of observed facts. By contrast, human beings, and their aspirations and values, could not be expelled from science. In turn, human beings and their wills, for her, are not simply determined by biological or sociological facts. Her insistence on human will does not mean that she was more conciliatory with pragmatism. In fact, she sketched a chapter dedicated to this philosophy that for her belittles human beings and their search for the truth. Human will is important to her, but it cannot be the only source of norms, as she interpreted pragmatists to hold. In her view, science, which she valued so deeply, was degraded by pragmatism to be the 'servant of industry', and human beings to be just producers (Metzger, 1954, pp. 57–59).

Above I cited Metzger writing that the loss of the capacity to distinguish between truth and error is fertile ground for totalitarianism. Her remark reminds the reader of Hannah Arendt's analysis, when she wrote that '[t]he ideal subject of totalitarian rule is ... the people for whom the distinction between fact and fiction ... and the distinction between true and false ... no longer exist' (Arendt, 2017 [1951], p. 622). Both Metzger and Arendt had direct experience of totalitarianism, and, although to my knowledge unknown to each other, both lived in Paris between 1933 and 1941. Metzger thought that the ideological use of science was partly responsible for people's apparent inability to think independently. For her, positivism, determinism, pragmatism, and the then current trends of biology and sociology left human beings without room for their aspirations to a different future. Similarly, Arendt wrote that ideologies offer 'total explanation', not just of current reality but also of change. She denounced the 'strict logic' used in totalitarian regimes, generally in the form of deductions from premises that are accepted as axioms, and that proceed 'with a consistency that exists nowhere in the realm of reality' (Arendt, 2017 [1951], p. 619). All this leaves no space for what Arendt called natality and new beginnings, nor for politics, which for her is the acting in concert of distinct human beings, and their collective creation of something new. She wrote that '[u]nderlying the Nazis' belief in race laws as the expression of the law of nature in man, is Darwin's idea of man as the product of a natural development which does not necessarily stop with the present species of human beings' (Arendt, 2017 [1951], p. 608). The future is mapped out, and does not depend on human desires, aspirations and actions. This is also the kernel of Metzger's critique of some philosophies and sciences of her time.

The central theme of Metzger's last work is the tension between human will and science. Arendt wrote that totalitarianism 'dispenses with human will to action altogether and appeals to the craving for some insight into the law of movement according to which the terror functions and upon which, therefore, all private destinies depend' (Arendt, 2017 [1951], p. 615). Like Arendt, Metzger thought that the negation of the human will as independent of laws of nature or society supported the tyranny of her times. Against the tyranny of what she referred to as the so-called scientific method, and of theories that left not space for freedom and novelty, Metzger proposed a 'balance' between thought and action (Metzger, 1954, p. 40). She saw an important space for science in human life; indeed she spent her life writing about the history of science, not to mention that she held a science degree. However, she could not accept that science could determine human will, and predict human destiny, or that facts would replace values. What is the place of religion in Metzger's argument? In negative terms, she emphasised that positivism has made the 'human soul' indifferent to God (Metzger, 1954,

p. 15); the Great Being not being a transcendent God, but only humanity itself (Metzger, 1954, p. 19). The other philosophies and scientific theories that she targeted left no room for God, or explained religion as a purely social phenomenon, as sociology did. She thought that they had failed. She invited to orient one's life towards the good, towards God and wisdom (Metzger, 1954, p. 40). She ended her sketch with an invitation to look for remedies to the current state of affairs in philosophy and religious traditions. She invited people not to abandon '[humankind's] immense progress', but rather to use what human beings have achieved in order to find again the path of their 'normal destiny', which they had 'momentarily' abandoned (Metzger, 1954, p. 61). I am not sure whether her mention of God and religion marked a change in her views, perhaps following a reflection brought about by what she was experiencing. There is no doubt, though, that she had remained critical of philosophies and sciences that leave no space for novelty, emotions, aspirations, creativity and freedom.

## Notes

1 Merllié has tentatively proposed that one of the reasons why Lévy-Bruhl has been forgotten is that it has been difficult to locate his work in relation to Durkheim's (Merllié, 1989).
2 Among the major philosophers in Metzger's milieu for whom Lévy-Bruhl work was important, one should mention Léon Brunschvicg and Abel Rey, who differently but extensively used Lévy-Bruhl's concepts (Brunschvicg, 1912, Brunschvicg, 1922, Brunschvicg, 1932, Brunschvicg, 1934, Brunschvicg, 1945–53, Brunschvicg, 1947); (Rey, 1930, Rey, 1937, Rey et al., 1937). Émile Meyerson also engaged, although very critically, with the concepts of participation and primitive mentality (Meyerson, 1931). Gaston Bachelard originally reworked the concept of primitive mentality, or primitive mind (Bachelard, 1964 [1938], Bachelard, 2002 [1938]).
3 For the differences between Lévy-Bruhl and the Durkheimians, as perceived by themselves, see the discussion between Lévy-Bruhl and Mauss (Lévy-Bruhl et al., 1923). For the foundation of the Institut d'ethnologie at the University of Paris, see Lévy-Bruhl's first report (Lévy-Bruhl, 1926).
4 Lévy-Bruhl wrote his *Notebooks* in 1938 and 1939, but they were only published in 1947.
5 See Chapter 5.
6 Metzger's image stands in contrast to Cartesian imagery of secure foundations, which she must surely have known. Descartes famously compared his method to that of an architect who lays the foundations of the house he plans to build on 'firm soil' (Descartes, 1984, p. 366).
7 On Metzger's analysis of the role of religion in van Helmont's work, see Halleux (1990).
8 Metzger attributes her quotation to Newton's Letter IV to Bentley, but her quote appears to be from Letter III. Her quotation also seems to be a little far from the original, which can be seen at http://www.newtonproject.ox.ac.uk/view/t exts/normalized/THEM00256 (Original letter from Isaac Newton to Richard Bentley, dated 11 February 1692/3).
9 On the theme of Metzger on Newton and religion, see Blay (1990).

10 Cf. (Rousseau, 1984 [1775], p. 85). The writer Ferdinand Brunetière converted to Catholicism in 1895; see Brunetière (1895).
11 As nowadays Adam Morton, among others, argues: see Chapter 8.

## References

Arendt, H. 2017 [1951]. *The origins of totalitarianism*, S.l., Penguin.
Bachelard, G. 1964 [1938]. *The psychoanalysis of fire*, Boston, Beacon Press.
Bachelard, G. 2002 [1938]. *The formation of the scientific mind*, Manchester, Clinamen Press.
Blay, M. 1990. Léon Bloch et Hélène Metzger: La quête de la pensée newtonienne. *In*: Freudenthal, G. (ed.) *Études sur / Studies on Hélène Metzger*, Leiden, Brill.
Boutroux, E. 1916 [1874]. *The contingency of the laws of nature*, Chicago, The Open Court Publishing Co.
Brunetière, F. 1895. *La science et la religion, réponse à quelques objections*, Paris, Firmin-Didot.
Brunschvicg, L. 1912. *Les étapes de la philosophie mathématique*, Paris, Alcan.
Brunschvicg, L. 1922. *L'expérience humaine et la causalité physique*, Paris, Alcan.
Brunschvicg, L. 1932. Nouvelles études sur l'anime primitive. *Revue des deux mondes*, 52, 172–202.
Brunschvicg, L. 1934. *Les âges de l'intelligence*, Paris, Alcan.
Brunschvicg, L. 1945–53. *Héritage de mots, héritage d'idées*, Paris, Presses Universitaires de France.
Brunschvicg, L. 1947. *L'esprit européen*, s.l., La Baconnière.
Burtt, E. A. 2003 [1924]. *The metaphysical foundations of modern science*, Mineola, NY, Dover, Newton Abbot, David & Charles.
Descartes, R. 1984. *The philosophical writings of Descartes*, Cambridge, Cambridge University Press.
Durkheim, É. 1909–12. I., Lévy-Bruhl, *Les fonctions mentales dans les sociétés inférieures*; É. Durkheim, *Les formes élémentaires de la vie religieuse. Année sociologique*, 12, 33–37.
Freudenthal, G. (ed.) 1990. *Études sur / Studies on Hélène Metzger*, Leiden, Brill.
Halleux, R. 1990. Visages de Van Helmont, depuis Hélène Metzger jusqu'à Walter Pagel. *In*: Freudenthal, G. (ed.) *Études sur / Studies on Hélène Metzger*, Leiden, Brill.
Lévy-Bruhl, L. 1884. *L'idée de responsabilité*, Paris, Hachette.
Lévy-Bruhl, L. 1890. *L'Allemagne depuis Leibniz. Essai sur le développement de la conscience nationale en Allemagne, 1700–1848*, Paris, Hachette.
Lévy-Bruhl, L. 1894. *La philosophie de Jacobi*, Paris, Alcan.
Lévy-Bruhl, L. (ed.) 1899a. *Correspondance de John Stuart Mill et d'Auguste Comte*, Paris, Alcan.
Lévy-Bruhl, L. 1899b. *History of modern philosophy in France*, London, Kegan Paul, Trench, Trübner & Co.
Lévy-Bruhl, L. 1903. *La morale et la science des mœurs*, Paris, Alcan.
Lévy-Bruhl, L. 1903 [1900]. *The philosophy of Auguste Comte*, London, Swan Sonnenschein & Co.
Lévy-Bruhl, L. 1905 [1903]. *Ethics and moral science*, London, Archibald Constable.

Lévy-Bruhl, L. 1910. *Les fonctions mentales dans les sociétés inférieures*, Paris, Alcan.

Lévy-Bruhl, L. 1924a. *Jean Jaurès. Essai biographique. Nouvelle édition suivie de lettres inédites*, Paris, Rieder.

Lévy-Bruhl, L. 1924b. L'idéal républicain. *Revue de Paris*, 31, 805–822.

Lévy-Bruhl, L. 1925 [1922]. *La mentalité primitive*, Paris, Alcan.

Lévy-Bruhl, L. 1926. L'institut d'ethnologie de l'Université de Paris. *Annales de l'Université de Paris*, 1, 205–209.

Lévy-Bruhl, L. 1963 [1931]. *Le surnaturel et la nature dans la mentalité primitive*, Paris, Presses Universitaires de France.

Lévy-Bruhl, L. 1975 [1947]. *The notebooks on primitive mentality, with a Preface by Maurice Leenhardt*, Oxford, Blackwell.

Lévy-Bruhl, L. 1996 [1927]. *L'âme primitive*, Paris, Presses Universitaires de France.

Lévy-Bruhl, L. 2015 [1910]. *How Natives Think*, Ravenio books, kindle edition.

Lévy-Bruhl, L. & et al. 1923. La mentalité primitive. Séance du 15 février 1923. *Bulletin de la Société française de philosophie*, 23, 17–48.

Merllié, D. 1989. Lévy-Bruhl et Durkheim. Notes biographiques en marge d'une correspondence. *Revue philosophique*, 114.

Merllié, D. 1993. Les rapports entre la *Revue de métaphysique* et la *Revue philosophique*: Xavier Léon, Théodule Ribot, Lucien Lévy-Bruhl. *Revue de métaphysique et de morale*, 98, 59–108.

Metzger, H. 1926. *Les concepts scientifiques*, Paris, Alcan.

Metzger, H. 1927. Lucien Lévy-Bruhl, *L'âme primitive*. *Isis*, 9, 482–486.

Metzger, H. 1938. *Attraction universelle et religion naturelle chez quelques commentateurs anglais de Newton*, Paris, Hermann.

Metzger, H. 1954. *La science, l'appel de la religion et la volonté humaine*, Paris, Boccard.

Metzger, H. 1969 [1923]. *Les doctrines chimiques en France du début du XVII^e à la fin du XVIII^e siècle*, Paris, Blanchard.

Metzger, H. 1974 [1930]. *Newton, Stahl, Boerhaave et la doctrine chimique*, Paris, Blanchard.

Metzger, H. 1987 [1930]. La philosophie de Lévy-Bruhl et l'histoire des sciences. *In*: Metzger, H. 1987. *La méthode philosophique en histoire des sciences. Textes 1914–1939, réunis par Gad Freudenthal*, Paris, Fayard.

Metzger, H. 1991 [1930]. *Chemistry*, West Cornwall, CT, Lucust Hill Press.

Metzger, H. 2019 [1936]. The *a priori* in scientific theory and the history of science (Appendix 3).

Meyerson, É. 1931. *Du cheminement de la pensée*, Paris, Alcan.

Rey, A. 1930. *La science dans l'antiquité, vol.1: La science orientale avant le grecs*, Paris, La Renaissance du livre.

Rey, A. 1937. L'évolution de la pensée: De la pensée primitive à la pensée actuelle. *In*: Febvre, L. (ed.) *Encyclopédie française*, Paris.

Rey, A., Meillet, A. & Monteil, P. 1937. *Encyclopédie Française: L'outillage mental*, Paris, Société de Gestion de l'Encyclopédie Française.

Rousseau, J.-J. 1984 [1775]. *A discourse on inequality*, London, Penguin.

Santillana, G. D. 1940. Hélène Metzger, *Attraction universelle et religion naturelle chez quelques commentateurs anglais de Newton*. *Isis*, 32, 145–148.

# Part III

# Metzger's legacy

# 7   Metzger's impact on her world

## An eternal pupil, or a teacher and an authority on the history of chemistry?

Although many critics have emphasised Metzger's originality, she is often presented in a manner that is common for female intellectuals. Even in excellent analyses of her work, the reader is often informed about the scholars who 'influenced' her, whereas her impact on other scholars is hardly ever discussed. Despite her repeated affirmations of independence from Meyerson, and her public declaration that she had already written two monographs before meeting him, she has been described as 'influenced by Meyerson and her own temperament' (Sarton, 1927, p. 468), one of Meyerson's 'disciples' (Bensaude-Vincent and Telkès-Klein, 2016, pp. 8, 257), indeed one of his 'young disciples' (Bensaude-Vincent and Telkès-Klein, 2016, p. 199). Discussions of the similarities and differences of her views and Meyerson's are framed as comments about her 'debt' to 'her mentor' Meyerson (Golinski, 1987), and her philosophy of science is seen as 'decisively influenced' by Meyerson's (Heidelberger, 1990, p. 151). Similarly, her original take on Lévy-Bruhl's concept of primitive mentality has been introduced, at least at first, as her application to the history of science of 'her uncle' Lucien Lévy-Bruhl's categories (Golinski, 1987); (Tosi, 2000).[1] She has been presented as 'Mieli's assistant' (Beretta, 2011, p. 271), and as 'Koyré's student', a description that, as we shall see below, though not factually incorrect, is very misleading indeed, especially when part of a one-paragraph biography that only provides the most essential information (Meyerson, 2009, p. 501). One could almost have the impression that she was never anybody's teacher, and that nobody read her works, until Thomas Kuhn discovered them. In reality, as we have seen, she was active and had posts of responsibility in the top institutions dedicated to the history of science. Indeed, Metzger was at the very centre of the development of history of science as a discipline both in France and internationally. She taught the very first students who could gain qualifications in this discipline in France, at the Sorbonne, where she also actively contributed to the organisation of teaching (Rey, 1933, p. 225); (Rey, 1935, p. 343); (Rey, 1936, p. 341); (Rey, 1937, p. 344). Her lectures at the École pratique des hautes études only covered one academic year, but she introduced history of science into the teaching of history of religion in an original way: rather than focusing on the impact of religion on science, as most scholars did, including Koyré, she looked at the impact of science on religion. She was even involved with the planning of the establishment of the history of science as an academic discipline in France.[2] As the secretary of the Unit for the history of science of the Centre de synthèse, she signed the published report of its activities that contains the text of a letter to be sent to the Ministry of Public Education. This letter was a petition for the creation of institutes of history of science in universities, as well as for the inclusion of history of science in the training of secondary-school teachers (Metzger, 1931); (Blay, 1997). The Sorbonne Institute was founded just the following year, with a full programme of courses and qualifications (Braunstein, 2006); (Chimisso, 2001). I do not mention the fact that Metzger wrote and signed the Unit for the history of science's report only in Berthold Brecht's spirit, when, in his poem 'Questions of a worker who reads', he imagined a worker who asks if really Alexander the Great conquered India on his own, as nobody else is mentioned, if splendid cities were really built, heavy rock by heavy rock, by kings, and, who, when informed of great military victories, wonders who cooked the feast for the victors.[3] Metzger certainly cooked many metaphorical meals, which were as necessary as Brecht suggested; for instance, as Simon Friederich Bodenheimer wrote to Sarton in 1950, the library catalogue of the International Academy for the history of science had been interrupted by Metzger's death (Bodenheimer, 1950, Letter to Sarton of 11.9.1950). Rather, I aim to emphasise Metzger's involvement in the important decisions regarding the new discipline of history of science. From her correspondence, it is clear that her roles at the International Academy and the Centre de synthèse were more important than her official titles as administrator and treasurer might suggest. Incidentally, the very name of 'International Academy for the History of Science' is due to Metzger, who successfully proposed the change from 'International Committee for the History of Science' (Comité International d'Histoire des Sciences, 1932).[4] Her service to the dissemination of history and philosophy of science is also clear when we see the volume of her reviews. Unsurprisingly, the largest number of them, over eighty, were published in *Archeion*, the organ of the Unit for the history of science of the Centre de synthèse; she also analysed new books for one of the two most important philosophy journals in France, *Revue philosophique*, and for *Thalès*, the journal of the Institut d'histoire des sciences. Notably, she introduced French philosophers of science to the American academic community of historians and philosophers of science thanks to her reviews for *Isis*.

As she is often introduced as somebody's mentee or student, it is easy to miss that in fact she was an authority in early-modern chemistry. French historians and philosophers of science referred routinely to her work. Gaston Bachelard, for instance, quoted Homberg, Boerhaave, Henckel and

Newton's commentator Cheney all indirectly via Metzger's *Les doctrines chimiques* (Metzger, 1969 [1923]), *Newton, Stahl Boerhaave* (Metzger, 1974 [1930]), and her articles and encyclopaedia entries (Bachelard, 1933, pp. 52–53); (Bachelard, 1973 [1932], pp. 20, 44); (Bachelard, 1984 [1934], pp. 73–74). However, he did not consider her work only as a handy shortcut to original sources, but also referred to her as the authority on alchemy and Paracelsian philosophy. In particular, as I shall discuss below, he referred to her regarding the importance of initiation in alchemy, and the role of analogy in early-modern thought (Bachelard, 2002 [1938], pp. 58, 95–96). That she was regarded as an authority on analogical thought is reinforced by other citations. A good example is the *Traité de logique et morale* that Georges Canguilhem published together with Camille Plenet in 1939.[5] When the authors explained analogical inferences, they referred to Metzger's *Les concepts scientifiques* (Canguilhem and Planet, 2011 [1939], p. 733). There is no doubt that Canguilhem read Metzger's works carefully. As mentioned, the work that came to his mind when he contrasted the object of science and the object of history of science was Metzger's *La genèse de la science des cristaux* (Canguilhem, 1994 [1966], pp. 16–17). In the same article, he also cited a past scholar, René Just Haüy, through Metzger.

It is more difficult to establish Alexandre Koyré and Hélène Metzger's mutual influences, although they were close in age, and they had greater contact than she had with Bachelard; Canguilhem belonged to a younger academic generation altogether. As Redondi has written, Koyré found his vocation as a historian of science in Paris (Redondi, 1983, p. 324). The Russian scholar, who had studied in Göttingen with Husserl and Hilbert[6] before moving to Paris, had a publication record that more easily fitted with history of philosophy, and indeed in Paris founded the philosophical journal *Recherches philosophiques*. However, he saw himself as a historian of ideas,[7] and he moved from the history of religious ideas to the history of scientific ideas. Koyré was undoubtedly a brilliant scholar, but in inter-war Paris he occupied a marginal place, as Canguilhem put it (Canguilhem, 1987). Metzger attended his lectures every year between 1934 and 1939, at first as a regular free auditor, then as 'graduate student'. The report also registered her 'active part' in the discussions of the lectures on Galileo (Koyré, 1986, pp. 43ff). It is, therefore, strictly speaking correct to say that she was Koyré's student, as the source mentioned above claims. However, to present her as such is very misleading. When she attended his lectures, she had had her *doctorat d'Université* for sixteen years; in fact, Koyré had obtained the same doctorate three years after her. She was an experienced historian of science, with five monographs to her name, and she had been lecturing at the Sorbonne's Institut d'histoire des sciences for years. On the one hand, she participated in his lectures, gained a diploma, and then she lectured there, when Koyré chose her, alongside Alexandre Kojève, to replace him in 1937–38, while he was teaching in Cairo.[8] On the other hand, when Koyré and Metzger met, he was working on his first book-length work in history of science, *Études galiléennes*, translated as *Galileo studies* (Koyré, 1978 [1939]). She tried to facilitate Koyré's membership of the community of historians of science by inviting him to give a paper at the Centre de synthèse, as well as proposing him for election as a member of the Unit for history of science (Section d'histoire des sciences, 1935). Rather than each other's students, Koyré and Metzger shared teachers and milieus, including Meyerson and his circle, Lévy-Bruhl and Brunschvicg.[9] We do not find Metzger's work discussed in Koyré's books, as we do in Bachelard's. However, in private, and many years after her death, he wrote to one of his students, who was writing a thesis on the history of chemistry, that he was ignorant of this discipline, and that he became interested in the development of science after Newton only thanks to Metzger's *Newton, Stahl, Boerhaave* (Zambelli, 2016, p. 248). In the next section, I shall discuss the position of Metzger's work in the context of historical epistemology, regarding in particular the conception of the scientific and historiographical objects. I will also point out Kuhn's ideas and approaches that are consistent with Metzger's lesson. There, I will have the occasion to briefly draw some contrasts between Metzger's and Koyré's views of science.

## In defence of impurity: Metzger and historical epistemology

The writer and chemist Primo Levi told how, while at university, he considered switching from chemistry to physics. It was 1941, and it seemed that Germany was going to win the war. Metzger was in Lyon writing *La science, l'appel de la religion et de la volonté humaine* (Metzger, 1954). Like her, Levi, a secular Jew, was in his way trying to find some answers in religion: he and others met in a Talmud Torah school, and sought 'the strength that overcomes injustice' in reading the Bible together. However, the massacres in the Polish ghettos brought home to him that he and his fellow Jews had no 'allies', either 'on earth' or 'in heaven' (Levi, 1985 [1975], p. 52). Up to that point, chemistry had been a source of certainty; chemistry for him had meant to go to the heart of Matter, and Matter, he wrote, was their ally because Spirit was dear to Fascism. However, now he was becoming critical: chemistry was empirical, 'had no theorems', and had ambiguous origins. The alchemists appeared to the young Levi to be confused in their ideas and language, interested in profit, gold in particular, and similar to magicians with their 'Levantine' tricks. By contrast 'at the origins of physics lay the strenuous clarity of the West – Archimedes and Euclid' (Levi, 1985 [1975], p. 53). The racism around him, which he apparently had partly absorbed, threw together chemistry and him as a Jew: they were united in having dubious and impure origins and practices. The young Levi, in any case, met a junior lecturer in physics who suggested to him that physics is in fact only aimed at giving order to phenomena, and does not go beyond them. The narration proceeds with Levi back to studying chemistry, trying to obtain pure benzene by distillation; he described his work as a 'ritual', and 'almost a religious act' (Levi, 1985 [1975], p. 60). In this instance, purity was not obtained, and Levi caused a spectacular explosion. He did not mention alchemy in relation to the fire he caused, but the reader has the impression that he has gone back to chemistry's roots, to being a philosopher of fire. This story at once shows the chemist's search for purity and the disastrous consequences of the lack of purity. However, this is only one aspect, and not the most important, of chemistry in Levi's presentation. In another story, 'Zinc', he tells how he learned that 'tender and delicate zinc, so yielding to acids', 'tenaciously resists' their attacks when very pure. Levi discarded one lesson that could be drawn from this story, namely the strength of purity. He very much preferred the opposite lesson: impurity wins, as it enables change and mutation, in short, life. Life, he wrote, needs impurities and 'impurities of impurities', whereas Fascism aimed at purity, and wanted everybody to be the same (Levi, 1985 [1975], p. 34). Levi's stories show to us chemistry as the science of impurity, both in practice and in its alchemical roots. At the same time, they also show that the chemist's work has often to do with purification, and with obtaining those pure substances, which, as Metzger stressed, were not available to early-modern scholars. Is chemistry the 'impure science', as for instance Bernadette Bensaude-Vincent and Jonathan Simon suggest (Bensaude-Vincent and Simon, 2012), or is it rather aimed at purification, material and epistemological, as for Bachelard?

As Bernadette Bensaude-Vincent has argued, chemistry has played a central role in shaping French philosophy of science (Bensaude-Vincent, 2005). Chemistry was crucial for Bachelard, Meyerson and Metzger, and was also important for Duhem. Historical epistemologists were fully conscious that their choice of a particular science shaped their epistemology. This is because, as we have seen in Chapter 5, for them epistemology is the *regional* study of the principles, methods and results of a science, based on the knowledge of the history of that particular science (Canguilhem, 2015 [1965]); (Canguilhem, 1994 [1968]), pp. 21–23); (Canguilhem, 2015 [1959]); (Bachelard, 1986 [1949], Chapter 4). Bachelard's focus on chemistry played an important role, alongside his analysis of physics, in his revision of Brunschvicg's neo-Kantian idealism. Brunschvicg mainly focussed on mathematics; he also studied the history of physics, although Émile Meyerson was not wrong when he said that Brunschvicg's interpretation of physics was dominated by his understanding of mathematics (Meyerson in (Brunschvicg and al., 1921)). Unlike mathematics, chemistry is the science of the transformation of matter, and it does not take place only in the mind. In Bachelard's words, 'the chemist is a builder of molecules' (Bachelard, 1951, p. 79). It is chemistry that inspired his view of materialism, which he painstakingly opposed both to the philosophers' materialism 'without matter', and to the materialism of imagination that he investigated in his books about reverie. His materialism is the 'materialism of matter', indeed of the 'enormous plurality of different matters' (Bachelard, 1972 [1953], p. 4).[10]

When Bachelard talked about phenomena, he did not mean phenomena in the Kantian sense, but rather in the scientific sense of phenomena created in the laboratory (Bachelard, 1951, p. 5); hence his famous concept of 'phenomenotechnique' (Bachelard, 1991 [1934], p. 17); (Bachelard, 2002 [1938], p. 70); (Bachelard, 1951, p. 92). Creation of new material realities is a human and rational activity; his materialism is 'experimental', 'rational' and 'educated'. Rationality imparts order on the otherwise bewildering variety of matters (Bachelard, 1951); (Bachelard, 1973 [1932]). For Bachelard, scientific work is construction, and crucially for us here, purification. As we shall see directly, purity is at the core of Bachelard's view of chemistry, science and indeed objectivity and rationality.

Different scholars learned different lessons from chemistry and its history; a well-known case is that of Bachelard and Meyerson, who constructed rather different epistemologies out of a shared focus on chemistry. The differences between the conclusions that Bachelard and Metzger respectively drew from the study of chemistry are less known, as are their differences in the role they assigned to purity and impurity in chemistry, epistemology and history. Certainly, Bachelard cited Metzger in order to support his own epistemological and historiographical theses. For instance, he singled out *Les doctrines chimiques* as the history of chemistry that provides the best and most comprehensive account of Paracelsian analogies, and cited Metzger with regard to the analogies between metals and human organs (Bachelard, 2002 [1938], pp. 95–96). Metzger had spent much energy demonstrating that those analogies were not irrational. She did not deny that the imagination and emotions played a role in them, but she did not think that only a purely rational activity could be part of the history of science. Bachelard drew the opposite conclusions from the examination of the same past theories, indeed from Metzger's presentation and interpretation of them. For him, Paracelsian analogies that she so well described do not 'encourage any research to be made', indeed 'they *put thought to flight*' (Bachelard, 2002 [1938], p. 96). Similarly, he employed her analysis and evaluation of alchemy in order to show the alchemists' 'primitive mentality' (Bachelard, 2002 [1938], pp. 148–149) and 'pre-scientific mind' (Bachelard, 2002 [1938], p. 94). He interpreted her descriptions of alchemic theories as 'pre-scientific reveries' (Bachelard, 2002 [1938], p. 96). For him, the alchemists had only 'dreams', and dreams could not be passed on to others, and even less be taught (Bachelard, 1972 [1953], p. 5). Unlike Metzger, Bachelard demanded purity from scientific objects as well as from scientific minds. His materialism is a *rational materialism*, as the title of one of his book declares (Bachelard, 1972 [1953]). In science, matters (matter is plural for Bachelard) are always purified, and not only in a metaphorical manner. For him, chemistry could not be properly founded until chemical elements could be guaranteed to be sufficiently pure (Bachelard, 1972 [1953], pp. 8, 12). The impure substances on which early-modern scholars experimented for him were not properly formed scientific objects. The purity that Bachelard demanded of the substances on which the chemist experiments has two aspects. One, just mentioned, is the industrial purification of substances, so that chemists can work on pure, standardised, stable, and always identical substances. The other aspect is the purity that results from their rationalisation. Substances are pure because they are stripped, indeed constantly and progressively stripped, of imaginative and a-rational projections. Within the discussion of scientific materialism, Bachelard reiterates the 'complete separation of rational life and oneiric life' (Bachelard, 1972 [1953], p. 19). He expressed keen agreement with Metzger when she discussed the alchemists' 'passions' (Bachelard, 2002 [1938], pp. 148–149), their ambitions, and love for mysteries (Metzger, 1969 [1923], p. 102). However, for him the alchemists, who 'sought mystery' in their substances (Bachelard, 1972 [1953], p. 25), had not purified their minds of images and desires. He quoted Metzger explaining that for van Helmont, laboratory work was a moral initiation and preparation for spiritual enlightenment (Bachelard, 2002 [1938], p. 58); (Metzger, 1969 [1923], p. 174). However, for him, this was evidence that van Helmont's work was not scientific, indeed stood in opposition to that of modern scientists whose 'emotional lives are no longer mixed up with their scientific lives' (Bachelard, 2002 [1938], p. 58).

Bachelard did not completely neglect to mention his disagreement with Metzger. He gently and briefly wrote that he and Metzger gave a different evaluation of the natural philosophers' view that minerals are living beings. As discussed earlier in the present book, Metzger traced the history of classification of the natural world, including the phases in which there was no clear distinction between minerals, vegetable and animal kingdoms. For her, variations in the classifications of nature are part of the evolution of natural knowledge, which involves rational and non-rational elements. Bachelard, however, responded that the belief that metals are endowed with life has less to do with the intellect than Metzger thought; indeed, it is a wholly 'affective' phenomenon, which originates in our desires and imagination, rather than in reason. For him, the endowment of inorganic matters with life was in fact one of the epistemological obstacles that he studied in *La formation de l'esprit scientifique* (*The formation of the scientific mind*), namely the animistic intuition (Bachelard, 2002 [1938], p. 160). He thought that the animistic intuition is enduring, and indeed it can be found in more recent texts than those that Metzger studied; he concluded that 'the more recent the fault, the more grievous the sin' (Bachelard, 2002 [1938], p. 160). By contrast, Metzger regarded 'impurity' as playing a crucial and positive role in the development of science, rather than being an epistemological obstacle. I have discussed how she opposed those philosophies, like logical empiricism, which aimed to eliminate spontaneous thought, imagination and feelings from knowledge. As noted above, Primo Levi referred to the ambiguous origins of chemistry, in particular to the greedy alchemists who attempted to obtain gold from other metals. Bachelard aimed to purify chemistry from those origins. For him, far from being early chemists, alchemists were really misers who desired gold, and hated squandering. Just as realists, they were dominated by greed rather than rationality (Bachelard, 2002 [1938], pp. 148–149). On the contrary, Metzger thought that the study of matter was a practice that belonged to the history of science even at times when chemists, or indeed alchemists, worked on 'impure' objects, both as chemical substances with impurities, and as objects that are not entirely rationally constituted. Similarly, she did not think that the history of science should be 'purified', as Bachelard proposed, nor that her study of what he called lapsed history was a mere 'palaeontology' of a scientific mind that no longer exists (Bachelard, 1951, pp. 24–25). Unlike Bachelard, she did not regard her primary sources either as impurities in a history of progressive rationalisation, or as epistemological obstacles. For Bachelard, chemistry appears to be the science of purity, or of progressive purification, whereas for Metzger it is the science of impurity, in its objects, the minds of its practitioners, and indeed its history.

Bachelard and Canguilhem have been seen as the main representatives of historical epistemology, indeed as 'twins'.[11] However, if we look closely enough, Canguilhem's philosophy and historiography present far less purity than those of his predecessor on the chair of history and philosophy of science at the Sorbonne. As far as the scientific object is concerned, Canguilhem was explicit about this. However, unlike the difference that I have drawn between Bachelard and Metzger on this topic, the differences between Bachelard and Canguilhem lay in the different sciences on which they focussed. Canguilhem's study of the history of the life sciences, medicine and psychiatry led him to be less strict, so to speak, about what a proper scientific object is than Bachelard was. The objects of his sciences can never be truly rationalised and purified, for several reasons, notably their individuality. Chemical substances are certainly plural,[12] as they are not undifferentiated matter, but each of them is the same in every laboratory. Indeed, as mentioned above, Bachelard argued that chemistry really became a proper science only when the possibility of having pure and identical substances arose. By contrast, an individual of a species cannot be made identical to another individual of the same species. Also, and this is crucial, the environment in which the individual lives is not a laboratory, it is not controlled, rationalised and constructed. Variations among individuals of the same species are normal, and so are those that are considered anomalies. Canguilhem emphasised that an anomaly can even have a positive value, because it can allow an individual to adapt to a new environment, or an environment that has changed, and in so doing to survive. Living beings and their environments change, indeed everything that is living changes, including an organic tissue or an organ. Canguilhem presented the problem of the object of the life sciences with these words:

When we think of a scientific object, we think of a stable object, identical to itself... But life? Is it not evolution, variation of forms, invention of behaviors?

(Canguilhem 1991 [1966], p. 203)

Canguilhem's historiography exhibits an analogous impurity. He did not state his difference from Bachelard, to whom he always paid homage. However, his history of science is rather different from Bachelard's rectified and 'sanctioned' history of science, from which lapsed theories and practices are ejected (Bachelard, 1951, Chapter 1). Life appears to invade history as well: just as the object of the life sciences could not be fully rationalised, so could not history of science. As Camille Limoges has warned, Canguilhem's *La formation du concept de réflexe*, despite being dedicated to Bachelard and having a title that closely resembles that of Bachelard's *La formation de l'esprit scientifique*, is not an application of Bachelard's epistemology (Limoges, 2015). In this book, Canguilhem created a continuity between Willis' concept of reflex movement and the modern concept. Willis elaborated his concept within an imaginative theory of light, which in fact appeared to be dominated more by what Metzger regarded as analogical thinking, and which made large use of images. Bachelard would have seen an epistemological break between Willis' concept of reflex and the current one, which as all scientific concepts could only emerge by overcoming the imaginative approach that had previously prevented any objectivity. I have argued elsewhere that Canguilhem in fact elaborated the concept of scientific ideology in order to re-admit theories and practices into the history of science that Bachelard's historiography would have regarded as epistemological obstacles, or at the very least as irrelevant for the advancement of science (Chimisso, 2015). Canguilhem's impure epistemology certainly owes to his study of the life sciences and to their 'impurity' from the point of view of philosophies, like Bachelard's and before him Léon Brunschvicg's, that presented science as liberation from prejudice, subjectivity, emotions and the imagination and as progressive rationalisation. For Canguilhem too rationalisation was important, but its bar was far lower than Bachelard's, as he admitted into the history of science theories that were critical of received notions, and because of that for him were rightfully part of history of science, but at the same time included imaginative approaches. When such theories were far removed from modern science, he chose to call them 'scientific ideologies', which, he emphasised, were not anti-science, but indeed on the side of science (Canguilhem, 1988 [1970]).

A relevant case study for our discussion is ancient atomism. For Bachelard, the intuition of atoms as the smallest and indivisible components of matter is central to ancient atomism as well as common knowledge. Both are opposed to science. Precisely because they sprung out of natural, 'first' intuition about nature, for Bachelard Lucretius' and Democritus' atomistic theories do not belong to the history of science. They do not exhibit the rational, discursive (as opposed to intuitive) approach of science. So much so that in his early work on 'atomistic intuitions', in which he discussed ideas about atoms in a variety of times and contexts, he did not organise his material chronologically but rather thematically (e.g. realist atomism, criticist atomism). This is because for him there is no history of intuitions, as they belong to the pre-scientific and ahistorical realm. Although Bachelard's citations of Metzger in *Les intuitions atomistiques* are very positive (Bachelard, 1933, pp. 31, 52–53, 64–65), his aims and his conclusions are very different from hers. Bachelard judged past theories from the point of view of modern science, and modern science for him had demonstrated 'the illusory character of our first intuitions'. He had not yet put forward his concept of epistemological break, but concluded that book by writing that philosophy would have 'to fill the void' that separates 'naïve atomism from modern scientific atomism' (Bachelard, 1933, p. 160). He was going to develop his philosophy by rejecting any filling of the void, and indeed regarding intuitions as always illusory. By contrast, Canguilhem regarded ancient atomism as a scientific ideology, and very much as part of history of science, indeed as a progressive part of it. For him, '[t]o the antiscience of religion they [Democritus and Lucretius] opposed the antireligion of science' (Canguilhem 1988 [1970], p. 33). Canguilhem's view of which theories and concepts are part of history of science resembles much more closely Metzger's than Bachelard's. The life sciences showed to him that rational purity cannot be achieved; Metzger reached the same view while analysing chemistry and crystallography. There is no doubt that he thought that Metzger's primary sources were the rightful object of history, as he chose her work to exemplify what the object of science is. It is hard to say how much his reading of Metzger's work impacted on his view. But, despite the closeness of Bachelard and Canguilhem, it is clear that on the issue of the object of the history of science Canguilhem stood closer to Metzger.

The lesson that Metzger learned from chemistry and in general the study of matter is that impurity is the motor of change and advancement. The impurity of chemistry also translated in an 'impure' view of historiography, epistemology and hermeneutics. For her, the empirical, interested, perhaps confused alchemists have the right to a place in the history of chemistry.[13] Metzger also made room for intuition, emotions, and indeed 'human will' in scientific knowledge. Moreover, she thought that in order to understand past sources, the historians can also employ sympathy, especially when straightforward analysis and data gathering prove insufficient to the comprehension of texts. Conversely, her epistemology and hermeneutics informed the broad and liberal choice of theories and practices that she judged relevant to the history of science. For her, if researchers employ their imagination, or are self-interested, it does not follow that their works are completely irrational, and that they should be dismissed as fantasy.

## An interrupted bridge from her world to ours

Thomas Kuhn and Hélène Metzger might appear to belong to rather different eras: she anchored to the interwar period, he solidly belonging to our side of the Second World War, and an enduring philosophical presence. Notwithstanding the important watershed of the war, we could take another perspective. The first edition of *The structure of scientific revolutions* came out in 1962, a mere eighteen years after Metzger's death. Had she lived, she could well have read it, as she would have been seventy-three. For Kuhn, she was one of the 'recent scholars', alongside Alexandre Koyré, Émile Meyerson and Anneliese Maier, whose work pointed to a new direction in the philosophy and historiography of the sciences. Kuhn also warned that readers would find few references to the scholars whom he listed as the main inspiration of his book, although he was, as he wrote, 'indebted to them in more ways than I can now reconstruct or evaluate' (Kuhn, 1996 [1962], pp. vii–ix). Nevertheless, in *The structure of scientific revolutions*, we find references to Metzger's *Les doctrines chimiques* regarding particular developments of chemistry (Kuhn, 1996 [1962], pp. 41–48). In his 'History of Science' entry for the *International Encyclopedia of the Social Sciences* (1968), Kuhn also included Metzger's *Newton, Stahl, Boerhaave* in his list of the 'fine studies' in 'new internal historiography', alongside others, notably Koyré's *Galileo studies* and *La révolution astronomique* (Kuhn, 1977, pp. 111, 125). These studies for him show the then new attitude, which is no doubt Metzger's, to study past science from the point of view of the original sources (Kuhn, 1977, p. 110). Indeed, he acknowledged that 'this group [Koyré, Meyerson, Metzger and Maier] has shown what it was like to think scientifically in a period when the canons of scientific thought were very different from those current today' (Kuhn, 1996 [1962], p. vii). Kuhn's precepts for the interpretations of past sources might have looked novel to some of his Anglophone readers, but he knew the historians and philosophers who had already made them their own. For instance, it has been stressed that Kuhn urged readers of the works of important thinkers to look for absurdities and ask themselves how a sensible person could have written them (Barnes, 1982); (Kuhn, 1977, p. xii). Kuhn's answer was that readers should search for the reasons of those apparent absurdities, which then would cease to be so, and in fact would guide the interpretation of the rest of the text, which would very likely change meaning in light of this new understanding. Kuhn's recommended method as we know is exactly what Metzger pledged to practice in relation to her sources: rather than dismissing them as irrational or incoherent, she aimed to understand those texts that appeared recalcitrant to rational analysis. Historians for her could have a chance to understand those texts only if they read them by keeping in mind the assumptions, worldviews and dominant concepts of the time, and only if they grasp their authors' aims, ambitions and emotions. In Kuhn's reading her efforts were successful, as for him she had shown how past practitioners made sense of nature and of their own practices. Kuhn's second general interpretative assumption, in Barry Barnes' systematisation, is that the author of a source belongs to a certain culture, and therefore the historian must be vigilant not to attribute modern meanings to terms employed in the source. I do not need to reiterate how important this was for Metzger, as I have discussed the problem of anachronism in Chapter 1. Indeed, as just mentioned, Kuhn acknowledged the lesson of Metzger and her fellow historians, namely to read past texts historically. Incidentally, I think that Metzger's approach, namely, to understand not mainly her sources' authors, but above all the minds of their contemporary readers, is in fact more sophisticated than the lesson that Kuhn appeared to draw. Kuhn has also been hailed by many as the historian who changed the face of a hitherto generalised Whig tradition (Barnes, 1982, pp. 3–4); (Bird, 2000, p. 267). This judgement is less generous than Kuhn's, as he acknowledged previous historians, including Metzger. The latter's criticism of the concept of precursors, and of a history of science as made out of a chain of 'great men' and remarkable discoveries, could not fail to attract his attention.

It must be said that Metzger would have not liked to be listed as an 'internal' historian, an expression that she did not know, although the reasons for Kuhn to do so are clear. Metzger's books are largely based on analysis of primary sources. Her work is not primarily about the study of institutions, or the impact of scientific concepts on 'various aspects of Western thought', or again on the social role and setting of science in a homogeneous local area, that is the types of history that in Kuhn's presentation make up 'external history' (Kuhn, 1977, pp. 113–114). On the other hand, as we have seen in Chapter 4, she did not see science as evolving independently of other cultural and social phenomena. In her historical work, she insisted on the diversity of social roles and occupations of the scholars she studied, and on the difference that the creation of academies made to their work. As mentioned, Kuhn also listed Koyré as an internal historian. Several authors have contested, or partially amended, the view that Koyré was an internal historian, notably Elkana, (Elkana, 1987); Zambelli (Zambelli, 1994); (Zambelli, 1995); and Stump (Stump, 2001). They emphasise Koyré's 'unity of thought', and, in Zambelli's case, the 'collective representations' derived from Lévy-Bruhl. If we apply the same arguments to Metzger, the case for her not being an internal historian is even stronger than for Koyré, as she had a much broader view of the types of thought that could have an impact on science. Indeed, Melhado has contrasted her approach with Koyré's focus on 'a handful of major innovators', and her defence of 'exogenous factors' in the explanation of science with Koyré's rejection of them. He has concluded that it is 'Metzger, not Koyré, who possesses relevance for the disciplinary history of eighteenth-century science' (Melhado, 1990, p. 117).

Kuhn dedicated a section of his article on history of science to discussing the new historiographical trends that took the Merton theses seriously. I use the plural because Kuhn presented Merton's explanation of the 'special productiveness of seventeenth-century science' as twofold. On the one hand, Kuhn explained, Merton emphasised the amount of data and experience that scientists acquired thanks to the attention to the techniques of craftsmen and in general to practical knowledge. On the other hand, Merton linked the rise of Puritanism and its valuing of work and communion with God through nature with the remarkable development of science (Kuhn, 1977, pp. 115–118). The link between religion and science was something, Kuhn wrote, that the 'old historiographic tradition' had often denied. Although Koyré and Metzger rather obviously are not discussed in the section dedicated to the Merton theses, the importance of religion for science is an important theme for both of them. However, their respective ways of seeing this link, I would argue, differ significantly from each other. For Koyré, change in theological thought and in scientific thought cannot be separated. The transformation from a closed world to an infinite universe, which for Koyré was the core of the scientific revolution, was at the same time a theological transformation. In his words:

> The Infinite Universe of the new Cosmology, infinite in Duration as well as in Extension, in which eternal matter in accordance with eternal and necessary laws moves endlessly and aimlessly in eternal space, inherited all the ontological attributes of the Divinity. Yet only those – all the others the departed God took away with Him.
>
> (Koyré, 1958 [1957], p. 276)

When Koyré proposed the unity of human thought, he considered it only 'in its highest forms', that is to say philosophy, theology and science. This is rather different from Metzger. She certainly agreed on the type of unity that Koyré suggested; it would be enough to think of her discussion of Cartesianism and corpuscular theory. She too indicated Newton's conception of God as central to his physics, as discussed in Chapter 6. However, her 'unity' was much broader, and certainly less patrician, than Koyré's. Not only theology, but also everyday religious assumptions, intuitions, experiences and sentiments for her have an impact on the work of the scientist. In general, when she discussed the conceptual framework of a period, she did not see it as belonging to high culture alone. Moreover, she discussed the impact of science on religion in *Attraction universelle et religion naturelle chez quelques commentateurs anglais de Newton* (Metzger, 1938); arguably, she could have claimed a place in one the categories of external historians, according to Kuhn's presentation. However, it must be said that Kuhn very likely never read that book.

Regarding the respective merits of internal and external history, Kuhn wrote that 'in the early development of a new field ... social needs and values are a major determinant of the problems on which its practitioners concentrated' (Kuhn, 1977, p. 118). Metzger focussed on early science, or, as Kuhn would have put it, on the process of maturation of a science, and she saw it as inextricably intertwined with the values and needs of its times. What about mature science? In his 'History of Science' article, Kuhn judged it as highly independent; in his expression, its practitioners become 'a special subculture', and they constitute a close group of sort, which is 'insulated from the cultural milieu in which they live their extraprofessional lives' (Kuhn, 1977, p. 119). Would Metzger have agreed with Kuhn's verdict? She did not write enough about modern science for us to give a full answer, and in truth she did not aim to answer this question. She never regarded science as value-free, but could some values only belong to a particular subculture? I do not think that Metzger would have agreed to such exclusivity, although she would certainly have conceded that the professionalisation of the sciences would produce particular values. For her, emotions and values would always be part of science, and she never suggested that the scientists' emotions and values are independent of the wider society. Certainly she did not think that regarding the historian, whose mental *a prioris* do not proceed from the community of historians. That community, as we have seen in Chapter 3, for her is the resource that the individual historian can access in order to reflect on her *a prioris* and correct misinterpretations of past sources due to the projection of her *a prioris* on them.

The more patrician view of science that Koyré proposed in relation to Metzger's takes us to another issue that Kuhn discussed in his presentation of various historiographical models. This is the role of manual work, experimentation and experience in the elaboration of scientific theories. Merton had seen it as crucial, whereas, Kuhn wrote, the new historians claimed to have shown that the scientific revolution 'owed very little to new instruments, experiments or observations' (Kuhn, 1977, p. 116). Koyré was the clearest representative of the latter trend, the same Koyré who so much irritated Mieli by saying that Galileo never performed the experiments he described, nor did he need to.[14] Science for Koyré 'is essentially *theoria*, a search for truth, and ... as a result of this fact it has, and has always had, value as an end in itself, and an inherent and autonomous ... development' (Koyré, 1963, p. 856).[15] His emphasis on science as intellectual enterprise may recall Brunschvicg's work, although the latter did not apply it to 'science' in general, but rather to mathematics, and to a certain extent to physics. As discussed above, Bachelard with his concept of phenomenotechnique brought the importance of technology and material objects to the fore. The history of chemistry, as Metzger presented it, is neither only theory, nor the reserve of a learned leisured class, as Koyré also suggested. As I mentioned throughout this book, Metzger sought the birth of chemistry in the activities of medics, pharmacists and assorted practitioners, and acknowledged the practical use of their knowledge. Indeed, one of the problems with her sources was that medics and pharmacists did not write out full explanations as they guarded their secrets for commercial reasons (Metzger, 1991 [1930], p. 31).[16] For her it was crucial to understand what aims her scholars had, because these were varied, and often were not a disinterested search for truth. I do not think that this difference with Koyré wholly depends on the fact that he focussed on 'classic' sciences whereas Metzger focussed on a 'Baconian' science in formation. There is no doubt that the processes that lead to the formation of a discipline cannot be internal to a discipline that does not yet exist. However, there is no ground to think that she would have regarded modern science as essentially theoretical, when, as we know, she thought that the first thing to learn in order to understand seventeenth-century chemistry was the laboratories and tools of the time. Or that early-modern pharmacists would have been moved by commercial reasons, whereas modern chemistry would have been immune to the interests of the pharmaceutical industry. In fact, she knew of this link personally as she was born to a family of precious stones merchants, and she had studied crystallography.

Kuhn also included Metzger's work among those histories of individual sciences that for him marked a new trend, in opposition to the previous attempts at general history of science, from Bacon's and Comte's, to Tannery's and Sarton's. Kuhn emphasised that 'experience' has shown that 'sciences are not all of a piece' (Kuhn, 1977, p. 109). It is perhaps regrettable that Kuhn did not engage with Bachelard's and Canguilhem's works, which must have vindicated his ideas also from a philosophical point of view, notably with their view of regional rationalism (Bachelard, 1986 [1949], Chapter 7). The 'new historiography' for Kuhn not only turned to particular branches of knowledge, but also respected past disciplinary distinctions, rather than imposing current ones onto the past. This of course is apparent in Metzger's work, as she insisted, for instance, that when studying the work of past practitioners concerned with the transformation of matter, the historian should examine the work of pharmacists, medics, alchemists and assorted scholars and amateurs, as we have seen in the previous chapters. In other words, she knew that it would have been anachronistic to write a history of chemistry, or a history of crystallography, by forcing the current disciplinary identities of these sciences onto the past.

The reader of *The structure of scientific revolutions* who is also acquainted with Metzger's work would find in it many other familiar ideas and approaches. Kuhn's critical targets had also been hers, notably, as he put it, the image of science based the study of 'finished scientific achievements, as these are recorded in the classics and, more recently, textbooks from which each new scientific generation learns to practice its trade' (Kuhn, 1996 [1962], p. 1). Needless to say, this was also the type of history Metzger was countering when she focussed on 'science in the making', as seen particularly in Chapter 2. She was very reflective on the type of sources that she employed, as she was aware that some did not reveal practices in full in order to avoid competition as just mentioned, others, although aimed at teaching students, would not reveal to the modern reader what was obvious for a contemporary and what a pupil would tacitly learn in the laboratory. Kuhn pointed out that classics and textbooks are not good bases to understand science, because they have pedagogical and persuasive aims. Again, Metzger was very reflective about the aims of histories of sciences as well as of her own sources. She noticed that many histories of science were motivated by national pride, while others were aimed at proving philosophical theses, traditionally the idea of progress, and in her time particular views of the mind, as discussed. Had Metzger been alive to read *The structure of scientific revolutions*, no doubt she would have felt greatly vindicated when Kuhn criticised presentist history of science. He claimed that the latter presents the historians with two main tasks: the attribution and dating of scientific discoveries and the explanation of superstition and error that hindered the formation of current science (Kuhn, 1996 [1962], p. 2). She was in a milieu in which the attribution of discoveries was very central, especially at the Centre de synthèse, and its Unit of the history of science. She repeatedly criticised such practice, in which she might have felt involved by the cataloguing tasks assigned to her. As she tried to explain to Sarton (Metzger, 1923, Letter to Sarton of 21.8.1923), she fully rejected the historiography of 'great works by great men'; she rather endeavoured to write the history of the formation of chemistry by following the slow changes 'in the majority of minds' (Metzger, 1969 [1923], p. 9). Partly against other historical epistemologists, Metzger also never accepted to treat lapsed history as a receptacle of 'superstition and error that hindered the formation of current science', to use the words, quoted above, that Kuhn employed to describe presentist history of science. Gaston Bachelard did precisely that with his concept of epistemological obstacle, but this view preceded him, although arguably in a less sophisticated form. Metzger approached past sources with humility, and rather than writing them off as products of non-scientific attitudes, she worked hard in order to understand what type of assumptions and values could lie behind their claims. Kuhn's work could well have been the bridge between Metzger's historiography and current scholarship. It has not been so, and it is beyond the scope of the present work to investigate the reasons for this missed opportunity. In the next and concluding chapter, I shall propose some ways in which Metzger's work can be still employed nowadays.

### Notes

1 Although many scholars, including Brunschvicg, Bachelard, Koyré, Rey and Lucien Febvre made large use of Lévy-Bruhl's categories as well, commentators do not often mention their 'debt' to Lévy-Bruhl, or their personal links to him (which in some cases were strong). For instance, Roger Chartier emphasises that Febvre employed Lévy-Bruhl's ideas, but his presentation of Febvre confers on the latter full agency and autonomy (Chartier, 1988, pp. 22–24).

2 A chair of history of science had been established in 1892 at the Collège de France (Paul, 1976). However, this particular institution does not offer qualifications. The foundation in 1932 of the Institut d'histoire des sciences at the University of Paris, with its full academic programme, marked more properly the birth of history of science as an academic discipline in France.

3 Here are some verses of Brecht's poem 'Questions of a worker who reads':

> Who built Thebes of the 7 gates?
> In the books you will read the names of kings.
> Did the kings haul up the lumps of rock?
> And Babylon, many times demolished,
> Who raised it up so many times?
> …
> Great Rome is full of triumphal arches.
> Who erected them?
> …
> The young Alexander conquered India.
> Was he alone?
> Caesar defeated the Gauls.
> Did he not even have a cook with him?
> …
> Every page a victory.
> Who cooked the feast for the victors?
>
> (Brecht, 1987, pp. 252–253).

4 I have given a flavour of the importance of some of her correspondence in the Introduction.

5 The *Traité* was a work for *lycée* pupils. Canguilhem and his co-author were consciously avoiding the word 'Manual' for pedagogical reasons: see Canguilhem and Planet (2011 [1939]), pp. 605–608.

6 (Delorme, 1965); (Koyré, 1986). He was part of Husserl's 'Göttingen group', which discussed the applications of phenomenology (Jorland, 1981, pp. 27–42); (Zambelli, 1999); (Kleinberg, 2005, pp. 27–28); (Schuhmann, 1987).

7 See Redondi in (Koyré, 1986, p. x).

8 Koyré held courses in the francophone Université égyptienne of Cairo on several occasions: 1933–34, 1936–37, and in 1940–41 (Zambelli, 2016).

9 Koyré and Lévy-Bruhl, despite their different political leanings, were very well acquainted; the latter convinced the former to translate the Schwartzkoppen papers, which proved captain Alfred Dreyfus' innocence (Schwartzkoppen, 1930); (Zambelli, 1995).

10 I have discussed the importance of chemistry for Bachelard's philosophy in (Chimisso, 2014).

11 An anonymous referee of one of my articles wrote that, unlike most of the secondary sources, I rightly did not regard Bachelard and Canguilhem as 'twins'.

12 Bachelard wrote extensively on chemistry's 'pluralism' (Bachelard, 1973 [1932]).

13 In her time, many scholars did not accept the inclusion of alchemy in the history of chemistry. For instance, in a petty, though not overall negative, review of Metzger's *Chemistry*, the Sorbonne professor of philosophy and history of philosophy Albert Rivaud (Guigue, 1935, p. 24) suggested that she was not as well acquainted with seventeenth-century chemistry as with modern chemistry. His contention was that alchemy was 'discredited' in the eighteenth century with the exception of Germany. For him, she should not have included those 'fantasists' (Rivaud, 1930). An illustration of the pettiness of Rivaud's review is that he corrected Metzger's spelling of an author's name, Johann Joachim Becher, which in *La chimie* is spelt Beccher. In fact, both spellings were in circulation, notably Pierre Duhem employed the spelling 'Beccher', like Metzger (Duhem, 1916). Incidentally, Albert Rivaud was close to l'Action française, an organisation on the extreme right, strongly anti-Dreyfusard. During the war,

he was briefly minister of education under Pétain, and in 1944 was suspended from his post at the Sorbonne for his role during the war (Charle, 1986, pp. 187–189).

14  Mieli disliked the talk that Koyré gave at the Unit for the history of science following Metzger's invitation (Section d'Histoire des Sciences, 1936). Although he did not speak at the talk, Mieli reserved his attacks for the printed page. For him the 'newcomer' Koyré had joined the ranks of Galileo's enemies by suggesting that Galileo never performed the leaning tower of Pisa experiment (Mieli, 1938). Koyré responded in a footnote of his *Galileo studies*, in which he wrote that '[w]e had thought that we had expressed our admiration for the genius of Galileo sufficiently clearly to make any misapprehension impossible ...' and went on to cite Mieli's article (Koyré, 1978 [1939], p. 210). No doubt these exchanges must have left Metzger rather cold.

15  Koyré continued by saying that science displays an 'autonomous development', and therefore it can be understood by the historian only by 'the study of its own problems, its own history' (Koyré, 1963, p. 856). He wrote this in the context of his argument against social explanations of science; as he succinctly put it, '[T]he social structure of England in the seventeenth century cannot explain Newton' (Koyré, 1963, p. 855).

16  See Chapter 1.

# References

Bachelard, G. 1933. *Les intuitions atomistiques (Essai de classification)*, Paris, Boivin.

Bachelard, G. 1951. *L'activité rationaliste de la physique contemporaine*, Paris, Presses Universitaires de France.

Bachelard, G. 1972 [1953]. *Le matérialisme rationnel*, Paris, Presses Universitaires de France.

Bachelard, G. 1973 [1932]. *Le pluralisme cohérent de la chimie moderne*, Paris, Vrin.

Bachelard, G. 1984 [1934]. *The new scientific spirit*, Boston, Beacon Press.

Bachelard, G. 1986 [1949]. *Le rationalisme appliqué*, Paris, Presses Universitaires de France.

Bachelard, G. 1991 [1934]. *Le nouvel esprit scientifique*, Paris, Presses Universitaires de France.

Bachelard, G. 2002 [1938]. *The formation of the scientific mind*, Manchester, Clinamen Press.

Barnes, B. 1982. *T.S. Kuhn and social science*, London, Macmillan.

Bensaude-Vincent, B. 2005. Chemistry in the French tradition of philosophy of science: Duhem, Meyerson, Metzger and Bachelard. *Studies in History and Philosophy of Science*, 36, 627–648.

Bensaude-Vincent, B. & Simon, J. 2012. *Chemistry: The impure science*, London, Imperial College Press.

Bensaude-Vincent, B. & Telkès-Klein, E. 2016. *Les identités multiples d'Émile Meyerson*, Paris, Honoré Champion.

Beretta, M. 2011. The changing role of the historiography of chemistry in continental Europe since 1800. *Ambix*, 58, 257–276.

Bird, A. 2000. *Thomas Kuhn*, Chesham, Acumen.

Blay, M. 1997. Henri Berr et l'histoire des sciences. *In*: Biard, A., Bourel, D. & Brian, E. (eds.) *Henri Berr et la culture du XXᵉ siècle. Histoire, science et philosophie*, Paris, Albin Michel/Centre international de synthèse.

Bodenheimer, F. S. 1950. Letter to Sarton. *Sarton papers bMS Am 1803*, Cambridge, MA, Houghton Library.

Braunstein, J.-F. 2006. Abel Rey et les débuts de l'Institut d'histoire des science et techniques (1932–1940). *In*: Bitbol, M. & Gayon, J. (eds.) *L'épistémologie française, 1830–1970*, Paris, Presses Universitaires de France.

Brecht, B. 1987. *Poems 1913–1956*, London, Methuen.

Brunschvicg, L. & AL. 1921. L'intelligence est-elle capable de comprendre? Séance du 24 février 1921. *Bulletin de la Société française de philosophie*, 21, 33–67.

Canguilhem, G. 1987. Preface. *History and Technology*, 4, 7–10.

Canguilhem, G. 1988 [1970]. *What is scientific ideology? Ideology and rationality in the history of the life sciences*, Cambridge, MA, London, The MIT Press.

Canguilhem, G. 1991 [1966]. *The normal and the pathological*, New York, Zone Books.

Canguilhem, G. 1994 [1966]. L'objet de l'histoire des sciences. *Études d'histoire et de philosophie des sciences concernant les vivants et la vie*, Paris, Vrin.

Canguilhem, G. 1994 [1968]. *Études d'histoire et de philosophie des sciences concernant les vivants et la vie*, Paris, Vrin.

Canguilhem, G. 2015 [1959]. Histoire générale des sciences. *Résistance, philosophie biologique et histoire des sciences 1940–1965; Œuvres completes tome IV*, Paris, Vrin.

Canguilhem, G. 2015 [1965]. Philosophie et science. *Résistance, philosophie biologique et histoire des sciences 1940–1965; Œuvres completes tome IV*, Paris, Vrin.

Canguilhem, G. & Planet, C. 2011 [1939]. *Traité de logique et de morale. Georges Canguilhem, Écrits philosophiques et politiques 1926–1939; Œuvres complètes, tome I*, Paris, Vrin.

Charle, C. 1986. *Les professeurs de la Faculté des Lettres de Paris. Dictionnaire biographique 1909 1939*, Paris, Institut national de Recherche Pédagogique/Éditions du CNRS.

Chartier, R. 1988. *Cultural history. Between practices and representations*, Ithaca, Cornell University Press.

Chimisso, C. 2001. *Gaston Bachelard: Critic of science and the imagination*, London, Routledge.

Chimisso, C. 2014. A Matter of Substance? Gaston Bachelard and chemistry's philosophical lessons. *In*: Galavotti, M. C., Nemeth, E. & Stadler, F. (eds.) *European Philosophy of Science – Philosophy of Science in Europe and the Viennese Heritage*. Dordrecht: Springer.

Chimisso, C. 2015. Narrative and epistemology: Georges Canguilhem's concept of scientific ideology. *Studies in History and Philosophy of Science*, 54, 64–73.

Comité International d'histoire des Sciences. 1932. Séance privée du Comité (14 mai, après midi) à la Sorbonne. *Archeion*, 14, 468–473.

Delorme, S. 1965. Hommage à Alexandre Koyré. *Revue d'histoire des sciences*, 18, 129–139.

Duhem, P. M. M. 1916. *La Chimie est-elle une science française?*, Paris.

Elkana, Y. 1987. Alexandre Koyré: between the history of ideas and sociology of disembodied knowledge. *History and Technology*, 4, 115–148.

Golinski, J. V. 1987. Hélène Metzger and the interpretation of seventeenth century chemistry. *History of Science*, 25, 85–97.

Guigue, A. 1935. *La Faculté des Lettres de l'Université de Paris depuis sa fondation (17 mars 1808) jusqu'au 1ᵉʳ Janvier 1935*, Paris, Alcan.

Heidelberger, M. 1990. History of science and criticism of positivism: Émile Meyerson's and Hélène Metzger's views from a present-day perspective. *In*: Freudenthal, G. (ed.) *Études sur / Studies on Hélène Metzger*, Leiden, Brill.

Jorland, G. 1981. *La science dans la philosophie: les recherches épistémologiques d'Alexandre Koyré*, Paris, Gallimard.

Kleinberg, E. 2005. *Generation existential: Heidegger's philosophy in France 1927–1961*, Ithaca and London, Cornell University Press.

Koyré, A. 1958 [1957]. *From the closed world to the infinite universe*, New York, Harper.

Koyré, A. 1963. Commentary. *In*: Crombie, A. C. (ed.) *Scientific change. Historical studies in the intellectual, social and technical conditions for scientific discoveries and technical invention, from antiquity to the present*, London, Heinemann.

Koyré, A. 1978 [1939]. *Galileo studies*, Hassocks, The Harvest Press.

Koyré, A. 1986. *De la mystique à la science. Cours, conférences et documents 1922–1962, édité par Pietro Redondi*, Paris, Ed. de l'École des hautes études.

Kuhn, T. S. 1977. *The essential tension: Selected studies in scientific tradition and change*, Chicago, University of Chicago Press.

Kuhn, T. S. 1996 [1962]. *The structure of scientific revolutions*, Chicago, The University of Chicago Press.

Levi, P. 1985 [1975]. *The periodic table*, London, Abacus.

Limoges, C. 2015. Introduction: Philosophie biologique, histoire des sciences et interventions philosophiques. *In*: G. Canguilhem, *Résistance, philosophie biologique et histoire des sciences 1940–1965. Œuvrès complètes, tome IV. Textes présentés et annotés par Camille Limoges*, Paris, Vrin.

Melhado, E. M. 1990. Metzger, Kuhn, and eighteenth-century disciplinary history. *In*: Freudenthal, G. (ed.) *Études sur / Studies on Hélène Metzger*, Leiden, Brill.

Metzger, H. 1923. Letters to Sarton. *Sarton papers bMS Am 1803 (1032)*, Cambridge, MA, Houghton Library.

Metzger, H. 1931. Activité de la Section d'histoire des sciences en 1930–31. *Revue de synthèse*, 51, 192–193.

Metzger, H. 1938. *Attraction universelle et religion naturelle chez quelques commentateurs anglais de Newton*, Paris, Hermann.

Metzger, H. 1954. *La science, l'appel de la religion et la volonté humaine*, Paris, Boccard.

Metzger, H. 1969 [1923]. *Les doctrines chimiques en France du début du XVIIᵉ à la fin du XVIIIᵉ siècle*, Paris, Blanchard.

Metzger, H. 1974 [1930]. *Newton, Stahl, Boerhaave et la doctrine chimique*, Paris, Blanchard.

Metzger, H. 1991 [1930]. *Chemistry*, West Cornwall, CT, Lucust Hill Press.

Meyerson, É. 2009. *Lettres françaises. Éditées par Bernadette Bensaude-Vincent et Eva Telkes-Klein*, Paris, CNRS.

Mieli, A. 1938. Il tricentenario dei 'Discorsi' di Galileo Galilei. *Archeion*, 21, 193–297.

Paul, H. W. 1976. Scholarship and ideology: The chair of the general history of science at the Collège de France, 1892–1913. *Isis*, 67, 376–397.

Redondi, P. 1983. Science moderne et histoire des mentalités. La rencontre de Lucien Febvre, Robert Lenoble et Alexandre Koyré. *Revue de synthèse*, 104, 309–330.

Rey, A. 1933. Institut d'Histoire des sciences et des techniques: Rapport du Directeur sur son organisation et son activité. *Annales de l'Université de Paris*, 8.

Rey, A. 1935. Institut d'Histoire des sciences et des techniques: Rapport du Directeur sur son organisation et son activité. *Annales de l'Université de Paris*, 10, 343–347.

Rey, A. 1936. Institut d'Histoire des sciences et des techniques: Rapport du Directeur sur son organisation et son activité. *Annales de l'Université de Paris*, 11, 340–345.

Rey, A. 1937. Institut d'Histoire des sciences et des techniques: Rapport du Directeur sur son organisation et son activité. *Annales de l'Université de Paris*, 12, 343–348.

Rivaud, A. 1930. *La chimie*, par Hélène Metzger. *Revue critique d'histoire et de littérature*, 377–378.

Sarton, G. 1927. *Les concepts scientifiques* by Hélène Metzger. *Isis*, 9, 467–470.

Schuhmann, K. 1987. Alexandre Koyré et les phénoménologue allemands. *History and Technology*, 4, 149–165.

Schwartzkoppen, M. V. 1930. *Les carnets de Schwartzkoppen: la verité su Dreyfus, édité par Bernhard Schwartfeger et traduits sur le texte allemand par Alexandre Koyré; préface Lucien Lévy-Bruhl*, Paris, Rieder.

Section d'histoire des sciences. 1935. Séance du 23 Janvier 1935. *Archeion*, 17, 81–84.

Section d'histoire des sciences. 1936. Séance du 22 janvier 1936. *Archeion*, 18, 238–247.

Stump, J. B. 2001. History of science through Koyré's lenses. *Studies in History and Philosophy of Science Part A*, 32, 243–263.

Tosi, L. 2000. Hélène Metzger y la historia de la química. *Saber y Tiempo*, 9, 77–87.

Zambelli, P. 1994. Fenomenologia, sociologia e storia delle idee in Alexandre Koyré. *In*: Vinti, C. (ed.) *Alexandre Koyrè: L'avventura intellettuale*, Napoli, Edizioni Scientifiche Italiane.

Zambelli, P. 1995. Alexandre Koyré versus Lucien Lévy-Bruhl: From collective representations to paradigms of scientific thought. *Science in Context*, 8, 531–555.

Zambelli, P. 1999. Alexandre Koyré alla scuola di Husserl a Gottinga. *Giornale critico della filosofia italiana*, 78, 303–354.

Zambelli, P. 2016. *Alexandre Koyré in incognito*, Florence, Leo S. Olschki.

# 8　Metzger in our world

The debt that science studies owe to Thomas Kuhn has been amply acknowledged. Kuhn has been hailed 'perhaps the most influential' twentieth-century philosopher of science (Bird, 2004), although the reception of his work has been particularly important among sociologists, as Jan Golinski has remarked (Golinski, 1998, p. 14). Indeed, in Alexander Bird's words, '... the impact [of *The structure of scientific revolutions*] on the social sciences was enormous' (Bird, 2000, p. 267). By contrast, Kuhn's own debt to previous studies of science, including Metzger's, has attracted little attention. As I put it in Chapter 7, Kuhn's work has failed to be a bridge between us and the ideas of the historians and philosophers of science whom he singled out as fundamental to the writing of *The structure of scientific revolutions*. Within that group, only Koyré has truly reached international fame, in particular in the English-speaking world, but his success has been quite independent of Kuhn's reception.[1] In fact, when the link between the two scholars has been emphasised, it has been for polemical ends, notably by Steve Fuller. Fuller discusses the intellectual connection between Koyré and Kuhn in reference to Platonism, and claims that Kuhn admired 'Koyré's brand of intellectual history' (Fuller, 2000, p. 43). Fuller uses Kuhn's reception of Koyré as evidence of the former's view of science as independent of society, and as essentially conservative (Fuller, 2000, pp. 60–72). Yet, even in Fuller's book, aimed as it is at showing that 'Kuhn saw as much as he did because he stood on the shoulders of giants' (Fuller, 2000, p. xii), there is very little attention to the other intellectuals whom Kuhn studied when writing *The structure of scientific revolutions*. Any person would see further when standing on anybody's shoulders, let alone a giant's, and arguably scholars enabling this improved vision could have attracted more interest; if that had been the case, it may have turned out that their role has not been one of mere support. Kuhn was certainly not the first scholar to assert the importance of history for philosophy; Metzger's intellectual milieu described in the present book is a powerful example of a collective effort to historicise philosophy of science. It was in Paris that historical epistemology developed. Indeed, in the English-speaking world, historical epistemology has become all but synonymous with French philosophy of science, as exemplified by the entry 'French philosophy of science' of the *Routledge Encyclopedia of Philosophy* (Gutting, 1998), distinct from the entry simply titled 'Philosophy of science' (Worrall, 1998). Currently, the expression 'historical epistemology' has been claimed by a number of historians of science and historically oriented philosophers. Some members of this group are very aware of the French legacy, including Hans-Jörg Rheinberger, Arnold I. Davidson and Ian Hacking (Rheinberger, 2010a, Rheinberger, 2010b); (Rheinberger, 2005b, Rheinberger, 2005a); (Davidson, 1997, Davidson, 1998); (Hacking, 1999a). Regrettably, this legacy has generally been restricted to a small number of French philosophers. Moreover, the link back to what I have come to call 'classic' historical epistemology, i.e. the epistemologies and historiographies developed in France in the first decades of the twentieth century, is not important for all current scholars who describe what they do as historical epistemology. Lorraine Daston once admitted that she did not even know where the name came from (Daston, 1994, p. 238). Nevertheless, as I shall briefly suggest below, many points of contact can be found between Daston's and Metzger's respective research programmes.

For a variety of reasons, which have little to do with Metzger's work, and more with history, language, institutions and sexism, current scholars have neither included her writings in their reception of historical epistemology, nor studied them as one of the roots of Kuhn's ideas. However, some current scholars have recognised the value of her work, notably Gad Freudenthal, who has long been committed to the diffusion of Metzger's ideas and writings, and of course others, several of whom are cited in this book. Indeed, some critics have urged current scholars to take her historiography as model alongside Kuhn's, notably Evan M. Melhado (Melhado, 1990). Ilana Löwy has also followed the thread that links Kuhn back to Metzger, and to Ludwik Fleck (Löwy, 1990).[2] Charles B. Schmitt not only believes that the significance of Metzger's historiographical papers has been 'severely undervalued', but even proposes to make the reading of 'Should historians of science ...?' (Metzger, 2019 [1933]) required 'for everyone who wishes to undertake the study of the history of science' (Schmitt, 1990, pp. 24–25). Oscar Abadía-Moro has shown that 'Metzger's ideas can provide historians of science with a useful set of tools to reflect upon "presentism"' (Moro-Abadía, 2008). Yet, Metzger's work is still marginal both as an object of study of historians of philosophy and intellectual historians, and as a source of historiographical lessons and philosophical ideas. Historical epistemology would look rather different, and certainly richer, if in addition to the works of Bachelard and Canguilhem, we read her works seriously, alongside those of Brunschvicg, Meyerson, Rey, Lévy-Bruhl and the other philosophers and historians of their milieu. Moreover, I propose that looking back at Metzger's work could complicate and illuminate current reflections on historiography, and in particular on the historiography of the sciences. It goes without saying that I do not suggest taking her work as a wholesale model. After all, scholars hardly ever uncritically receive past philosophies and historiographies in their entirety. Although Bachelard has been taken as an inspiration by current historical epistemologists, his lesson, if taken in its entirety, would have led them to discard the whole of natural philosophy as an obstacle to science, and to dismiss the role of emotions and habits in science. We know that this has not been the case. Equally, it cannot be said that Kuhn's rich legacy has uniformly translated into works that have taken Kuhn's work as a rigid paradigm. As Alexander Bird has suggested, Kuhn's ideas 'have played little part' in the development that they helped trigger, including Bruno Latour and Steve Woolgar's social construc-tivism, and Karin Knorr Cetina and Sandra Harding's critiques of science (Bird 2000: 270). Indeed, the Edinburgh school, which arguably has devel-oped its Strong Programme of sociology of scientific knowledge in closest dialogue with Kuhn's work, has been, rightly or wrongly, disavowed by Kuhn himself in a late address (Kuhn, 1992); (Bird, 2004).

Metzger's work, when considered historically, inevitably belongs to its time in assumptions and aspirations. Her first book is now over a century old, and in that and her other works, she addressed questions that were urgent and exciting in her time. She wrote her history of science in conscious opposition to the then mainstream historiography, which was focussed on the results of science, rather than its processes, and on great men and discov-eries. Similarly, it was for her a priority to combat the dominant model of scientific development as displaying a clear progressive direction. In the face of a strongly presentist tradition in the history of science, she thought that it was imperative to insist on a properly historical understanding of past sources. Many current readers may cringe at her use of the term 'mentality', and her important references to the work of Lévy-Bruhl.[3] However, refer-ences to Lévy-Bruhl's ideas were common, indeed standard, among intel-lectuals who studied the mind, such as the *Annales* historians Lucien Febvre and Marc Bloch, sociologists like Marcel Mauss and Émile Durkheim, (Durkheim, 1902–3, Durkheim, 1909–12), psychologists like Henri Wallon (Wallon, 1928), and, as I have discussed, philosophers like Brunschvicg, Meyerson, Rey and Bachelard. Indeed, in the 1950s, Emmanuel Levinas traced Lévy-Bruhl's impact on a different philosophy from that discussed in the present book, namely existential philosophy (Levinas, 1957). At the same time, as I hope I have shown, an unprejudiced approach to the discus-sion of the so-called mentalities may in fact be still relevant and stimulating. Indeed, as I shall mention below, there have been recent calls for reconsid-eration of those ideas, notably from Mary Douglas. As other twentieth-century philosophies, Metzger's sits in an ambiguous position between our past and our present. Even philosophers who read past works without his-toricising them are aware of the distance that separates them from Aristotle and Descartes. Twentieth-century works generally appear to be of more immediate fruition, and it may be easier to forget that the contexts of their appearance, as well as the questions that they were aimed to answer, may differ quite significantly from our own. It is important to bear in mind that at the time of Metzger's writing history of science was a young discipline that was only then finding its own feet, and still needed to be defended. The reader may recall from the Introduction that Metzger's crystallography professor dismissed her manuscript of *La genèse de la science des cristaux* (Metzger, 1918) as being on a topic of no interest to anyone. Moreover, in France, history of science was institutionally linked to philosophy, largely practised by philosophers, and aimed at answering philosophical questions. The awareness of those local circumstances should help us to understand the works of the time, so that they can speak to us more clearly. On the other hand, as I have already pointed out, her works are closer in time to Kuhn's *Structure of scientific revolutions* than the latter work is to us. It is a balancing act to acknowledge the different aims and critical targets of works belonging to the interwar period, and at the same time not to read them as if they belonged to a far-removed era. As J. R. R. Christie puts it, '[Metzger] is still … an historian whom we wish either to argue against or enlist in our support, a present voice as much as a past text' (Christie, 1987, p. 108). Indeed, some of her aims are rather close to those for instance of current historical epistemologists, and the latter may find interesting ideas in her work. Lorraine Daston mentions two fundamental characteristics of her historical epistemology, which distinguish it from the history of ideas. First, historical epistemology asks 'the Kantian question about the preconditions that make thinking this or that idea possible'. This, as I have amply discussed, is central to Metzger's writings: her work on high-level concepts is a systematisation of the preconditions that made the appearances of ideas in the history of chemistry possible. The second characteristic of Daston's historical epistemology is that it 'radically challenges the assumption of resemblance between ideas advanced by thinkers working within different conceptual categories' (Daston, 1994, p. 282). Again, I do not think that there is need to expand on this, as I have amply discussed Metzger's criticism of anachronism and of the concept of precursors, as well as her insistence on the importance of understanding ideas and practices in the context of the conceptual, material and indeed emotional worlds in which they emerged. Are there other ideas that Metzger can still fruitfully offer to our reflection? I shall here make some brief proposals, which are not aimed at being exhaustive. Other readers will surely find other ideas in her writings that command their attention.

## Anachronism

I have dedicated Chapter 1 of the present book to an analysis of Metzger's sustained effort to make herself a contemporary of her sources, as she put it. My choice to employ the term 'anachronism' in order to name Metzger's critical target is not casual. She did not use this term, which however connects her to us, as it is in current use. In employing it, I also adopted a

Metzgerian attitude to terminology, which I shall briefly discuss below. Although for Metzger it was urgent to react to the 'triumphalist' history of science, which was inherently anachronistic, this issue has not gone away, and arguably never can. When we construct narratives, the risk of anachronism is always present. Do we misread past experiments, theories, social practices? Are we sufficiently aware of past practitioners' aims and questions? Are we reading the past proleptically? On the other hand, is a degree of presentism inevitable? These are fundamental questions of historiography. A good document that shows the endurance of these concerns is Quentin Skinner's 'Meaning and understanding in the history of ideas', originally published in 1969, and more recently republished as a chapter of his book *Visions of politics* (Skinner, 2002). Skinner's paper has been of great importance for the historiography of the sciences, although it addresses the history of political ideas. Skinner, like Metzger, warns that the same term may significantly change its meaning. If the historian is not alert to this, she may assume that a past source refers to something that in fact its author could not have known. 'To the great despair of historians, men fail to change their vocabulary every time they change their customs', pointed out Marc Bloch, followed by Ginzburg who quotes him (Bloch, 1992 [1954], p. 28); (Ginzburg, 2013c, p. 97). We could add that they fail to change their vocabulary every time they change theories of the natural world, and discover or create new substances. Metzger for instance pointed out that in the seventeenth century the term 'phosphorus' referred not only to a 'defined chemical body' but also to a variety of matters that give out a similar light (Metzger, 1969 [1923], p. 412). Similarly, the historian would err in assuming that the terms 'water' and 'ice' were taken by sixteenth-century scholars to refer to the same substance in different states, as they are now. In fact, as I noted in Chapter 1, those scholars believed that water was transformed into ice when mixed with 'frigorific' particles (Metzger, 1926, p. 70). Skinner also warns against reading past ideas as 'approximation to an ideal type', and therefore as 'anticipating' later doctrines, almost in a case of 'clairvoyance' (Skinner, 2002, p. 63). Again, it is difficult not to think of Metzger's attack on the concept of 'precursor', who for her is presented as a kind of 'prophet'. Although Metzger did not theorise the 'mythology of doctrines', in Skinner's terminology, she emphasised time and time again that many of her sources' authors did not have a coherent theory about this or that problem. For instance, she wrote that early-modern experimental chemists did not follow a clear theory, and sometimes did not quite grasp the theory that they claimed to follow (Metzger, 1969 [1923], p. 343). In fact, some of them were not interested in theory at all, as they were more concerned with, say, collecting and describing 'angular stones'.[4] Skinner and the historians of science who have followed his lesson, such as Andrew Cunningham, put great emphasis on the understanding of the categories that past agents employed in order to provide a historically accurate rendition of their ideas and practices. This is again very similar to Metzger;

indeed, she wrote *Les concepts scientifique*s in order to systematise the categories the use of which she detected in her sources.

This communality of concerns, however, should not lead us to believe that Metzger's historiographical ideas perfectly overlap with Skinner's or with those of the historians of science who follow his lead. Like Skinner, she was concerned with what past authors 'meant', but she was more interested in their contemporaries' understanding of their works. In this book I have insisted on this, for at least two reasons. One is that she very clearly focussed on collective representations, in line with the sociology and ethnology of her time. Her emphasis on authorship is far weaker than Skinner's. The second is that the 'meaning' of a text from a historical point of view is not given, or at least not entirely, by its author's intentions, as Cunningham among others argues. Her main preoccupation was how contemporary readers and pupils of a natural philosopher would, or could, have understood the latter's text or instructions. As mentioned in Chapter 3, she argued that it is precisely because different readers interpreted the same text in different ways that sometimes new research projects emerged, and innovations were introduced. From her perspective, authorial intentions are not necessarily the most relevant historically and scientifically. Her view seems to me close to Nick Jardine's, when he defends 'a reception-based view of meaning, taking the senses of historical sources to be grounded not in the intentions of their authors, but in the ways in which fully competent readers of the period would have understood them' (Jardine, 2019, p. 34). Other current historians would agree with Metzger; among them, Carlo Ginzburg has gone as far as to say that often the historian should read sources *against* the intentions of their authors (Ginzburg, 2019). Ginzburg's sources, namely the minutes of Inquisition trials, were constructed from the inquisitors' perspective, while the voices of the defendants were distorted and silenced (Ginzburg, 2013b, Ginzburg, 2013a). In order to gain a less biased picture of past events, Ginzburg has aimed to recover those voices as much as possible. Incidentally, his stance recalls that of Steven Shapin, who in a classic paper has aimed to re-establish the presence of 'invisible' technicians in the making of natural knowledge; the images that accompany his article well illustrate the ambiguous role of technicians whose individual identities were concealed from the viewer (Shapin, 1989). Metzger aimed to recover a number of voices, belonging to authors who were excluded from the history of science, to practitioners who did not author anything, but left traces in others' work, and to the pupils who absorbed their masters' teaching without leaving any documentation of that transmission of practices. She also aimed to recover ideas that played important roles in theories of matter but were not explicitly stated, let alone attributed to someone, as they were simply too widely accepted and taken for granted. She walked a tight rope between taking seriously the limits of what scholars in a certain time could think and do, and acknowledging their ability to modify or oppose received theories and practices, and indeed the possibility that some scholars could simply misunderstand them. Her approach, therefore, should be of interest to those scholars who on the one hand study science as a collective enterprise and focus on collective ideas and categories, and on the other regard ideas and practices as susceptible to variations at local and indeed individual level.

Metzger aimed to make herself a 'contemporary' of her own sources in order to translate them for current readers. Her goal was to make those texts understandable and relevant to her readers, who had rather different views of nature and science. *Traduttore traditore* [translator traitor]? Well, certainly to a certain extent, and inevitably so. A complete and unencumbered access to past sources would not only be impossible, but it would erase the work of the historian as a constructor of narratives. As a constructor of narratives, the historian has a very different aim from that of the authors and readers of early-modern texts about transformation of matter. Notably, as discussed in Chapter 5, Metzger assigned philosophical aims to the writing of the history of science, including that of understanding the human mind. It goes without saying that those aims did not belong to her sources' authors. Moreover, she did not aim to make her voice disappear in her work. Rather, she thought that the historian should engage in a dialogue with her sources; in this dialogue both past and present categories are inevitably and fruitfully at work. For instance, her concept of mental *a priori* was derived from Lévy-Bruhl's concept of mentality, which had entered the academic world not much earlier than Metzger's first works. This was a category that was obviously alien to her sources, and yet it played a key role in her interpretation of early-modern texts. Her conscious use of historical, philosophical and ethnographical categories of her time makes her approach close to that of Nick Jardine. He has argued that the use of 'categories alien to the agents studied is often perfectly legitimate', and that a variety of successful approaches do precisely that, including in his list '[h]istory of mentalities, perspectival history, gender history, critique of ideology, psychobiography, sociological history (whether moderate interest-theoretic à la Shapin, network-theoretic à la Latour, field-theoretic à la Bourdieu), history of material culture and the other new cultural histories' (Jardine, 2000, p. 265). Metzger's partial presentism has in fact attracted the attention of commentators, notably Oscar Moro Abadía (Moro Abadía, 2011, Moro Abadía, 2009, Moro Abadía, 2008). Metzger took a similar approach when it came to what I have named terminological anachronism. One the one hand, like Skinner, she signalled the errors that historians may commit when they assume that a term had the same referent in the past as it has in the present, as mentioned above. On the other hand, especially in her popular book *Chemistry*, she did not shrink from using anachronistic terminology, for instance when saying that Scheele and Priestley discovered 'oxygen' (Metzger, 1991 [1930], p. 62), although this term was coined by Lavoisier. She also occasionally 'translated' terminology in order to inform the modern reader of which substance her scholars were manipulating. Was she falling in the error that she herself warned against? Here the issue is philosophical, or more precisely ontological, rather than purely historiographical. Many historians and sociologists would disapprove of Metzger's practice. In order to illustrate Metzger's position, I shall use Bruno Latour's presentation of 'radically anti-whiggish history', to which incidentally he does not subscribe, at least not in its simple form. His own position is beyond the scope of my discussion, as is his self-confessed misrepresentation of the case he takes as an example. Latour considers the statement that the Egyptian Pharaoh Ramses II, who lived over 3,000 years ago, could have died of tuberculosis. 'The attribution of tuberculosis or of Koch's bacillus to Ramses II should strike us as an anachronism of the same calibre as if we had diagnosed his death as having been caused by a Marxist upheaval, or a machine gun, or a Wall Street crash'. This would be the case because 'Koch's bacillus [was] discovered (or invented, or made up, or socially constructed) in 1882' (Latour, 2000, p. 248). Latour's presentation might be somewhat extreme, but it does not misrepresent the view of historians who aim to describe the past fully according to the categories of the time. Ramses II could not have had the concept of tuberculosis or have discussed Koch's bacillus. The use of the term tuberculosis in this case is analogous to Metzger's use of the term oxygen when discussing Scheele, and her listing the names of the substances that early-modern scholars would have lumped into their 'phosphorous' category. Metzger's attitude strikes me as comparable to Ian Hacking's. The latter has analysed different 'styles of thinking and doing' in science, a concept that owes much to Michel Foucault in particular, but more in general to the French tradition of which Metzger was part (Sciortino, 2017). At the same time, he also holds that there are natural kinds, or more precisely in his terminology 'indifferent kinds'. These kinds are indifferent to the way we think about them and classify them, as opposed to 'interactive kinds', that is to classifications that can influence what they classify. 'Hyperactive child', 'criminal' or 'the homosexual' for him belong to the latter category, whereas plutonium, water, sulphur and horses are indifferent to human classifications (Hacking, 1999b, pp. 103–107). Metzger spent much energy in showing that classifications of natural objects greatly varied from one time to another. Nevertheless, she did not really doubt that the substances themselves were indifferent to the various classifications. For her, we can interact today with the substances with which alchemists interacted, provided that we find out to which substances the relevant sources referred. Had she not held such position, she would have not been able to assert the importance of re-enactment of past experiments (Metzger, 1969 [1923], p. 343) and in general of grasping the material environment in which early chemists operated (Metzger, 1991 [1930], Chapter 1). If we follow what she thought, it would not have been anachronistic for her to say that Koch's bacillus was responsible for Ramses II's demise, although it would have been trivially anachronistic to suggest that Ramses or his doctor thought that the bacillus was the cause of his illness.

## Science in the making, emotions and mentalities

The discussion about anachronism largely presupposes a historical view of science, and of science as human activity, or process, rather than a collection of truths, as Cunningham among others has pointed out (Cunningham, 1988). Ian Hacking has claimed that a main source of mutual incomprehension between sociologists and scientists arises precisely because they conceive of science differently: for the former it is a process and for the latter an assemblage of truths and a product (Hacking, 1999b). Metzger took as her object science as activity and process, or, as I put it in Chapter 2, 'science in the making'. The philosophers in her milieu aimed to overcome the distinction between science as process and as collections of truths: for historical epistemologists truth emerges from a historical process and particular human activities. Arguably Bachelard provided the clearest example of this effort of synthesis: for him it is a 'rationalist' and 'technical' activity that enables modern physics to produce truths. These truths, however, are historical, and they can be superseded if a more rationalist, and more technical, activity emerges (Bachelard, 1986 [1949]); (Bachelard, 1951); (Bachelard, 2002 [1938]). Metzger's focus on 'science in the making' is philosophically in line with the central perspective of classic historical epistemology. There are, however, important philosophical differences between her view of science in the making and those of philosophers like Brunschvicg and Bachelard, as I have discussed in Chapter 5. The latter had a normative view of the history of science: for them current theories provide the norms by which to judge past theories. They employed those norms to demarcate science from non-science and truth from falsehood in the history of science. Metzger was interested in understanding how current science had come about, and was not a relativist: she certainly did not think that the alchemists' project to transform any metal into gold was a good research programme. Nevertheless, in contrast with Brunschvicg's and Bachelard's, her approach was not normative. In her history writing, she avoided any judgement of past beliefs: she presented theories as true or false from the point of view of the natural philosophers whom she investigated. Indeed, with an attitude that may recall in current readers the Edinburgh school's symmetry principle, she refused to analyse differently successful and unsuccessful knowledge claims, to use David Bloor's language (Bloor, 1991). Moreover, she attached a cognitive value to theories and activities that from the point of view of current science were not scientific. She aimed to show that past ideas and practices that philosophers like Bachelard relegated to the rank of lapsed history of science in fact contributed to the development of science, sometimes in fostering critical attitudes, other times in opening up new fields of investigation, bringing new objects to the attention of scholars, developing good experimental practices, and indeed making important discoveries, such as the double refraction in crystals of Iceland spar that I mentioned in Chapter 2. As I have discussed, in this she stood in disagreement with the philosophers in her milieu, and for different reasons, with other contemporaries, including Karl Popper. Arguably, Metzger's focus is more similar to that of later historians, such as Gerald Holton, who aims to understand the 'nascent phase' of science (Holton, 1988).

Another crucial difference between Metzger and the philosophers close to her, notably Brunschvicg and Bachelard, is that she was also not normative about the *type* of process that science is. In particular, and against most of her peers, she did not think that the process needs to be fully rational. In fact, she insisted on the value of emotions, intuitions and sympathy both in the work of scientists and of historians, as I discussed in Chapter 3. Her approach raised many eyebrows in her milieu, but it should be of interest to many current historians and philosophers. One of the strongest current models is Daston's 'moral economies' of science, as 'balanced system[s] of emotional forces' (Daston, 1995, p. 4).[5] On my reading this shares much with Metzger's view that emotions are integral to scientific work. The latter did not explore emotions to the extent that Bachelard, repeatedly quoted in Daston's article, did. However, whereas Bachelard regarded emotions as obstacles to science, Metzger thought that they are a positive part of it, just as they are part of any type of knowledge. Daston is careful to distinguish her project from a psychological approach. In particular, she points out that moral economies are mental states not of individuals, but of collectives. It is redundant to repeat here that Metzger's approach is indeed that of considering collective rather than individual ways of thinking, doing and feeling. Indeed, it is this approach that steered her towards adopting the point of view of the pupils and readers of the scholars she studied rather than concentrating on authorial intentions, as noted above. She was well aware that her scholars' pupils learned from their teachers not only theories, practices and skills, but also ways of feeling, hence her use of sympathy in order to try to access this part of scientific training. Once again, this recalls Daston's view that '[a]pprenticeship into a science schools the neophyte into ways of feeling as well as into ways of seeing, manipulating, and understanding' (Daston, 1995, p. 5). Not only historians of science, but also many philosophers nowadays defend the role of emotions in the formation of knowledge. Adam Morton, among others, has shown the importance of epistemic emotions in the correct acquisition of beliefs. In his hypothetical case of a scientist who possesses many qualities and talents, but whose research is hindered by her lack of crucial emotions, he cites wonder. He illustrates this example by mentioning that his hypothetical scientist 'does not feel wonder at the connection between facts that she can glimpse through the data' (Morton, 2010). Lorraine Daston and Katharine Park have dedicated a voluminous monograph to wonder as a 'cognitive passion' with a long and important history. However, wonder, they write, '[s]ince the Enlightenment has become a disreputable passion in workaday science, redolent of the popular, the amateurish and the childish'; '[T]o be a member of the modern elite', they add, 'is to regard wonder and wonders with studied indifference' (Daston and Park, 1998, pp. 14–15, 368). Yet Morton cannot imagine a successful scientist who does not feel wonder. Do emotions play an enduring role in the formation of knowledge, notwithstanding the negative philosophical judgements passed on them at a time or another? Metzger did not investigate wonder in particular but analysed the roles of other emotions and passions in the formation of scientific as well as historical knowledge. She observed the same emotions and passions of her early-modern scholars in modern scientists, historians and general public. Doubtlessly, one of her historiographical concerns was to grasp the different epistemic role that emotions played in the past. However, she justified her attempts to access, however partially, the collective ways of feeling of past actors precisely with the enduring contribution to the same emotions in the formation of knowledge, notwithstanding the differences in their roles. Her defence of the role of sympathy and intuition in historical knowledge, which was unfavourably received by the historians of the Centre de synthèse (see Chapters 3 and 5), could present current historians with interesting suggestions. Indeed, Nick Jardine's defence of an empathetic engagement with past lives and works appears to be rather close to Metzger's view, also because he shares with her a reception-based view of meaning, with a focus on the receptions contemporary to the sources (Jardine, 2019).

Against Enlightenment and positivist traditions, Metzger re-admitted into science not only emotions, but also ways of thinking that in the eyes of her peers fell short of modern rationality, as discussed in Chapter 6. In her milieu, her view was not mainstream: Lévy-Bruhl led the way by counterpoising modern mentality to the 'primitive' mentality of non-European cultures. Historians and philosophers of science found 'primitive' ways of thinking in the past of European cultures, and Bachelard detected primitive mentality not only in the past, but also in the poetic and dreaming self of modern people. Most philosophers and historians with whom Metzger was in contact regarded scientific and 'primitive' ways of thinking as opposed to each other; Meyerson, who did not, regarded them as fundamentally the same. Metzger detected different ways of thinking in her sources, but did not think that they were mutually exclusive: for her, just as alchemists were not strangers to rational thought, so scientists and modern historians are not strangers to ways of thinking that Lévy-Bruhl and other philosophers associated with 'primitivity'. Many, including Lévy-Bruhl and Bachelard, agreed that analogy was at the core of primitive mentality. Metzger originally based her own concept of active analogy on Lévy-Bruhl's concept of participation, which was for him the hallmark of primitive mentality. However, as we have seen, she presented active analogy as a fundamental concept not only of Renaissance medicine and alchemy, but also of Newtonian and indeed current science. In the period that separates us from Metzger, the interest in different ways of thinking, epistemes, paradigms and styles of thinking has taken so many forms and played so many different roles that it would be impossible to discuss them here. However, it is perhaps worth noting that this interest has generally refrained from 'rehabilitating' Lévy-Bruhl. Yet, Mary Douglas in 2007 wrote that it was time to revisit the works of Lévy-Bruhl, who, as she wrote, may have provided wrong answers, but had posed the right questions. These questions regard the fundamental modes of thinking, which, Douglas proposed, are two: analytical (or 'hypothetic-deductive, predicative, rational-instrumental, discursive'), and analogical ('primitive, magico-religious, mytho-poetic') (Douglas, 2007). She backed up her views on the fundamental role of analogical thinking with Pierre Bourdieu's *The logic of practice* (Bourdieu, 1990 [1980]), among other works, and claimed that the role of analogy in science no longer needs to be demonstrated after such works as Mary Hesse's *The structure of scientific inference* (Hesse, 1974), Holyoak and Thagard's *Mental leaps: Analogy in creative thought* (Holyoak and Thagard, 1995), and Keane's *Creativity, analogy and consciousness* (Keane, 1991). This is in contrast with Lévy-Bruhl's times, in which, she claimed, it was 'impossible to imagine' other ways of knowing the world besides providing the correct evidence (Douglas, 2007, p. 8). I propose here that Metzger in that time could imagine another way, and that her work would be a good point of reference, arguably better than Lévy-Bruhl's. Unlike him, she did not ascribe different ways of thinking to different civilisations; she also wrote extensively on the role of analogical thinking, which she subdivided in different types, as analysed in Chapter 1.

On the one hand, Metzger took differences in doing and thinking in different times very seriously. On the other, by defending the enduring role of emotions, analogical thinking and religion in modern science, she renounced the main criteria of demarcation between science and non-science employed by philosophers since the Enlightenment, and notably in her milieu by Brunschvicg and Bachelard. In this, she could be a point of reference for many current scholars. I have cited but a few examples of scholars who have studied the role of emotions and analogical thinking in the formation of knowledge. Another norm of demarcation between science and non-science has also been secularism; Metzger's teacher Brunschvicg made it a specific theme of his work, as seen in Chapter 5, but it goes without saying that the opposition of religion and science has been mainstream since the Enlightenment. Metzger had a nuanced view of the relationship between religion and science; notably she analysed both the impact of religion on science and that of science on religion, as discussed in Chapter 6. Her detailed studies on the mutual influences of religion and science could be of interest to some current scholars. Here I shall mention only an example among many: Peter Dear would have found ammunition in Metzger's writing when criticising Andrew Cunningham's 'demarcation' between natural philosophy and science based on the centrality of God in the former enterprise (Dear, 2001). Other demarcation criteria, in Sandra Harding's words, have 'withered away' in the last decades. Harding mentions the traditional demarcation criteria, including 'a distinctive method', 'a critical attitude towards traditional belief, a distinctive language (mathematics, observation

sentences), a distinctive metaphysics (disenchanted, secular, material ...)'
(Harding, 2015, p. x). In this context, Metzger's work is very interesting as
she showed in her histories and argued in her papers that those criteria do
not hold. As we know, Metzger showed the varieties of methods, practices,
metaphysics and belief systems that contributed to chemistry and crystallography. Indeed, as mentioned in Chapter 5, she reacted rather strongly to the
programme of logical empiricism because of what she saw as its attempt to
create a language that would deliver objectivity. Metzger's refusal to oppose
in an absolute manner rationality and emotions, discursive and analogical
thinking, science and religion produced a model of development of scientific knowledge that does not fit with the usual representations of French
historiography as an opposition between 'continuists' and 'discontinuists',
as I have argued in Chapter 4. In fact, her non-linear view of the history of
science, which, by borrowing her expression from another context, I have
named 'theme and variations', may still be of interest to current historians. These scattered and very partial suggestions are aimed to show that
Metzger's work can be highly relevant to a number of current discussions.
However, they are not meant to advance the absurd claim that Metzger
'anticipated' ideas and priorities of later philosophers, historians and sociologists. Rather, adopting an approach that Metzger would recognise, I have
aimed to give a historically accurate rendition of her historiography in the
context in which it was developed. At the same time, I have selected themes
and ideas that in my view can still be relevant today, although inevitably
in new contexts and very likely in revised form, just like the ideas of any
past scholar. The point of writing any history, this being history of science,
history of philosophy, intellectual history or social history, in the end is to
serve our present concerns.

## Notes

1 Koyré's fame was supported by the contingencies of history; whereas Metzger
 was not able to leave France in the face of Nazi invasion, Koyré's move to the
 United States created a new audience for his ideas and yet a new language for his
 writings.
2 Löwy interprets Metzger's and Fleck's epistemologies as contructivist, and incidentally argues that their similar scientific training was at least in part responsible for them (Löwy, 1990).
3 For a number of years, I have been an invited speaker at a graduate seminar in the
 Department of History of and Philosophy of Science, University of Cambridge.
 Geoffrey Lloyd, who on that occasion organised the series with Nick Jardine,
 was clearly happier with Metzger's interest in the practice of science and criticism of anachronism than with her use of the concept of mentality. However, I
 think – but of course I might be wrong – that Metzger's version of mental *a priori*
 appeared to him more palatable than Lévy-Bruhl's 'mentality'. He has presented
 his critical view of mentalities in Lloyd (1990).
4 It goes without saying that Skinner's and Metzger's sources are rather different.
 The sources of the historian of political ideas may certainly have, and often do
 have, practical dimensions and aims; it suffices to think of Machiavelli or Marx.

However, a historian of chemistry like Metzger reads descriptions of experiments and observations, handbooks that teach how to perform particular transformations, alongside more theoretical treatises.
5 For an application of Daston's moral economies of science, see Bensaude-Vincent (2016).

# References

Bachelard, G. 1951. *L'activité rationaliste de la physique contemporaine*, Paris, Presses Universitaires de France.

Bachelard, G. 1986 [1949]. *Le rationalisme appliqué*, Paris, Presses Universitaires de France.

Bachelard, G. 2002 [1938]. *The formation of the scientific mind*, Manchester, Clinamen Press.

Bensaude-Vincent, B. 2016. The moral economy of synthetic biology. *In*: Boldt, J. (ed.) *Synthetic biology, metaphors, worldviews, ethics and law*, Wiesbaden, Springer.

Bird, A. 2000. *Thomas Kuhn*, Chesham, Acumen.

Bird, A. 2004. Thomas Kuhn. *In*: Zalta, E. N. (ed.) *The Stanford Encyclopedia of Philosophy*, Stanford, The Metaphysics Research Lab.

Bloch, M. 1992 [1954]. *The historian's craft*, Manchester University Press.

Bloor, D. 1991. *Knowledge and social imagery*, Chicago, London, University of Chicago Press.

Bourdieu, P. 1990 [1980]. *The logic of practice*, Cambridge, Polity.

Christie, J. R. R. 1987. Narrative and rhetoric in Hélène Metzger's historiography of eighteenth century chemistry. *History of Science*, 25, 99–109.

Cunningham, A. 1988. Getting the game right: some plain words on the identity and invention of science. *Studies in History and Philosophy of Science*, 19, 365–389.

Daston, L. 1994. Historical Epistemology. *In*: Chandler, J., Davidson, A.I. & Harootunian,H. (eds.) *Questions of evidence: Proof, practice and persuasion across disciplines*, Chicago, University of Chicago Press.

Daston, L. 1995. The moral economy of science. *Osiris*, 10, 2–24.

Daston, L. J. & Park, K. 1998. *Wonders and the order of nature, 1150–1750*, New York, Zone Books.

Davidson, A. I. 1997. *Foucault and his interlocutors*, Chicago, University of Chicago Press.

Davidson, A. I. 1998. *The emergence of sexuality: Historical epistemology and the formation of concepts*, Cambridge, MA, Harvard University Press.

Dear, P. 2001. Religion, science and natural philosophy: thoughts on Cunningham's thesis. *Studies in History and Philosophy of Science*, 32, 377–386.

Douglas, M. 2007. Raisonnements circulaires: retour nostalgique à Lévy-Bruhl. *Sociological Research Online*, 12.

Durkheim, É. 1902–3. Lévy-Bruhl, *La morale et la science des mœurs. Année sociologique*, 7, 380–384.

Durkheim, É. 1909–12. L. Lévy-Bruhl, *Les fonctions mentales dans les sociétés inférieures*; É. Durkheim, *Les formes élémentaires de la vie religieuse. Année sociologique*, 12, 33–37.

Fuller, S. 2000. *Thomas Kuhn: a philosophical history for our times*, Chicago, University of Chicago Press.

Ginzburg, C. 2013a. *Clues, myths, and the historical method*, Baltimore, MD, Johns Hopkins University Press.

Ginzburg, C. 2013b. *The night battles: Witchcraft and agrarian cults in the sixteenth and seventeenth centuries*, Baltimore, The Johns Hopkins University Press.

Ginzburg, C. 2013c. Our words, and theirs: A reflection on the historian's craft, today. *Cromohs*, 18, 97–114.

Ginzburg, C. 2019. Carlo Ginzburg presenta *Nondimanco. Talk at the Università degli Studi di Milano*, 6 April 2019.

Golinski, J. V. 1998. *Making natural knowledge: Constructivism and the history of science*, Chicago, University of Chicago Press.

Gutting, G. 1998. French philosophy of science. *In*: Craig, E. (ed.) *Routledge encyclopedia of Philosophy*, London, Routledge. Retrieved 9 October 2006 from http://www.rep.routledge.com/article/Q038.

Hacking, I. 1999a. Historical meta-epistemology. *In*: Carl, W. & Daston, L. (eds.) *Wahrheit und Geschicte*, Göttingen, Vandenhoeck & Ruprecht.

Hacking, I. 1999b. *The social construction of what?*, Cambridge, MA, Harvard University Press.

Harding, S. G. 2015. *Objectivity and diversity: Another logic of scientific research*, Chicago, The University of Chicago Press.

Hesse, M. B. 1974. *The structure of scientific inference*, London, Macmillan.

Holton, G. 1988. *Thematic origins of scientific thought: Kepler to Einstein*, Cambrige, MA, Harvard University Press.

Holyoak, K. J. & Thagard, P. 1995. *Mental leaps: Analogy in creative thought*, Cambridge, MA, London, MIT Press.

Jardine, N. 2000. Uses and abuses of anachronism in the history of the sciences. *History of Science*, 38, 252–270.

Jardine, N. 2019. Emotional engagement in scientific biographies. *In*: Matravers, D. & Waldow, A. (eds.) *Philosophical perspectives on empathy: Theoretical approaches and emerging challenges*, Abingdon, Routledge.

Keane, M. T. 1991. *Creativity, analogy & consciousness*, Dublin, Trinity College, Department of Computer Science.

Kuhn, T. S. 1992. *The trouble with the historical philosophy of science*, Robert and Maurine Rothschild dinstinguished lecture, 19 November 1991, An occasional publication of the Department of the History of Science, Cambridge, MA, Harvard University Press.

Latour, B. 2000. On the partial existence of existing *and* nonexisting objects. *In*: Daston, L. (ed.) *Biographies of scientific objects*.

Levinas, E. 1957. Lévy-Bruhl et la philosophie contemporaine. *Revue philosophique*, 82, 556–569.

Lloyd, G. E. R. 1990. *Demystifying mentalities*, Cambridge, Cambridge University Press.

Löwy, I. 1990. Constructivist epistemologies: Metzger and Fleck. *In*: Freudenthal, G. (ed.) *Études sur / Studies on Hélène Metzger*, Leiden, Brill.

Melhado, E. M. 1990. Metzger, Kuhn, and eighteenth-century disciplinary history. *In*: Freudenthal, G. (ed.) *Études sur / Studies on Hélène Metzger*, Leiden, Brill.

Metzger, H. 1918. *La genèse de la science des cristaux*, Paris, Alcan.

Metzger, H. 1926. *Les concepts scientifiques*, Paris, Alcan.

Metzger, H. 1969 [1923]. *Les doctrines chimiques en France du début du XVIIᵉ à la fin du XVIIIᵉ siècle*, Paris, Blanchard.

Metzger, H. 1991 [1930]. *Chemistry*, West Cornwall, CT, Lucust Hill Press.

Metzger, H. 2019 [1933]. Should historians of science become the contemporaries of the scholars they study? (Appendix 1).

Moro Abadía, O. 2008. Beyond the whig history interpretation of history: Lessons on 'presentism' from Hélène Metzger. *Studies in History and Philosophy of Science*, 39, 194–201.

Moro Abadía, O. 2009. Thinking about 'presentism' from a historian's perspective: Herbert Butterfield and Hélène Metzger. *History of Science*, 47, 57–77.

Moro-Abadía, O. 2011. Hermeneutical contributions to the history of science: Gadamer on 'presentism'. *Studies in History and Philosophy of Science*, 42, 372–380.

Morton, A. 2010. Epistemic emotions. *In*: Goldie, P. (ed.) *The Oxford handbook of philosophy of emotion*, Oxford, Oxford University Press.

Rheinberger, H.-J. 2005a. Gaston Bachelard and the notion of 'Phenomenotechnique'. *Perspectives on Science*, 13, 313–328.

Rheinberger, H.-J. 2005b. Reassessing the historical epistemology of Georges Canguilhem. *In*: Gutting, G. (ed.) *Continental philosophy of science*, Oxford, Blackwell.

Rheinberger, H.-J. 2010a. *On historicizing epistemology: An essay*, Stanford, Stanford University Press.

Rheinberger, H.-J. 2010b. On the historicity of scientific knowledge: Ludwik Fleck, Gaston Bachelard, Edmund Husserl. *In*: Hyder, D. & Rheinberger, H.-J. (eds.) *Science and the life-world: Essays on Husserl's 'Crisis of European sciences'*, Stanford, Stanford University Press.

Schmitt, C. B. 1990. Some considerations on the study of the history of seventeenth-century science: Lessons from Hélène Metzger. *In*: Freudenthal, G. (ed.) *Études sur / Studies on Hélène Metzger*, Leiden, Brill.

Sciortino, L. 2017. On Ian Hacking's notion of style of reasoning. *Erkenntnis*, 82.

Shapin, S. 1989. The invisible technician. *American Scientist*, 77, 554–563.

Skinner, Q. 2002. *Visions of politics*, Cambridge, Cambridge University Press.

Wallon, H. 1928. La mentalité primitive et celle de l'enfant. *Revue philosophique*, 53, 82–105.

Worrall, J. 1998. Science, philosophy of. *In*: Craig, E. (ed.) *Routledge encyclopedia of philosophy*, London, Routledge. Retrieved 3 December 2012 from http://www.rep.routledge.com/article/Q120).

# Appendices

## Four talks by Hélène Metzger

Hélène Metzger gave the talks translated into English in these Appendices at the Centre de synthèse, between 1933 and 1937. It is very important to bear in mind that these were talks rather than articles, although they were published at the time in the journal *Archeion*. Their style is very different from the works that Metzger prepared for publication. They lack references and bibliography, and, more importantly, their style is that of spoken rather than written language. Moreover, their tone is polemical, and at times defensive. She presented her historiographical ideas to a largely hostile audience. The director of the Centre, Henri Berr; the director of the Centre's Unit for the history of science, Aldo Mieli; and other historians who attended her talks held a positivistic view of history that Metzger comprehensively criticised. These talks were part of larger debates; this is clear in the discussions that followed her talks (see below for details) and from her anticipations of objections to her claims. Also, at the beginning of 'The philosophical method in the history of science', Metzger says that the topic of her talk was chosen by Henri Berr. We can presume that they had already discussed aspects of the historical method, and that Berr invited her to expand on her claims, which were at odds with his own. Moreover, Metzger used these talks to respond to criticism, even when the critic could not be in the audience, as in the case of Meyerson, who had already died at the time of her talk 'The philosophical method'. As I remark in the main part of the present book, in these talks she presented her views in a rather clear-cut, and sometimes exaggerated, manner. For instance, she appears to dismiss the importance of the study of the material part of science, when in fact she paid keen attention to it. This is understandable as she was speaking to an audience who dismissed her focus on the study of science in the making and spontaneous thought, and who greatly valued the accumulation of verifiable facts and events. In short, these talks are best read with her polemical aims in mind. They are of great value, both because in them she presented her historiographical and philosophical ideas in a straightforward manner, and because they give us access to the historiographical debates in which her work developed. Metzger aimed to study science in the making; her talks give us access to history of science in the making.

I have kept Metzger's original use of masculine pronouns to indicate a person of either gender. This is not only because it was standard use at the time, but also because I do not wish to hide the pervasive sexism of the milieu and culture in which Metzger lived and worked.

I have translated the French term *savant* sometimes as 'scholar' and sometimes as 'scientist', depending on the context. 'Scientist' may sound anachronistic in places to the modern reader, but Metzger employed the term 'science' for early-modern natural philosophy. I have also kept her chemical metaphors, even when they might sound odd to some readers.

The original talks have been re-published in Hélène Metzger, *La méthode philosophique en histoire des sciences, textes 1914–1939*, edited by Gad Freudenthal, Paris, Fayard, 1987.

*Archeion* also published the discussions that followed Metzger's talks:

1. *Archeion*, 15 (1933), 154–59: discussion of Metzger, 'Should historians of science become the contemporaries of the scholars they study?' [first published as 'L'historien des sciences doit-il se faire le contemporain des savants dont il parle?', *Archeion*, 15 (1933) 34–44];
2. *Archeion*, 17 (1935), 82–84: discussion of Metzger, 'The tribunal of history and the theory of scientific knowledge' [first published as 'Tribunal de l'histoire et théorie de la connaissance scientifique', *Archeion*, 17 (1935), 1–14];
3. *Archeion*, 18 (1936), 75–79: discussion of Metzger, 'The *a priori* in scientific theory and in the history of science' [first published as 'L'*a priori* dans la doctrine scientifique et l'histoire des sciences', *Archeion*, 18 (1936), 29–42];
4. *Archeion*, 19 (1937), 254–57: discussion of Metzger, 'The philosophical method in the history of science' [first published as 'La méthode philosophique dans l'histoire des sciences'. *Archeion*, 19 (1937), 204–216].

# Appendix 1: Should historians of science become the contemporaries of the scholars they study?[1]

I invited you all here today to discuss a topic that is of the greatest interest to me. I would like to examine the following question with you: 'Should historians of science attempt to make themselves contemporaries of the scientists whose theories they expound?' This is a question of method that I encountered as soon as I set out to study the evolution of scientific thought. I have already had the opportunity to answer that if the historian does not want to, or even know how to, read past works as the first readers would have done, he would run the risk of misunderstanding them, and as a result not only of misrepresenting the works of our predecessors, but also of altering the perspective of scientific progress itself ... This response has not been entirely well received by those who take an interest in historical method. I have been reproached for refusing to judge past theories with the hindsight of our current theories. It has been said that it was right to celebrate some past chemists for thinking about a particular issue about which we still think today. It has been said that one could reproach some past chemists for being so far removed from the way of seeing that has become our own ... You will appreciate that I am committed to justifying my point of view, since, in fact, several historians of science, who have not troubled themselves with historical criticism, have (after superficially acquainting themselves with the texts) produced accounts that resemble a prize list; [they have done so] by constantly referring to a positivist notion of science, or to the particular theory accepted at the time of publication of their accounts.

You can clearly see that this problem, upon which I am urging you to debate openly, is one of the most important, perhaps even the greatest methodological problem in history of science. This is because the chosen solution may determine our entire notion of humanity's past, and of the role of human thought, experience, positive empiricism or metaphysical inspiration in the very formation of science. Conversely, as I am sure you will reply, it is our preconceived notion of humanity's past, the role of human thought, experience, positive empiricism or metaphysical inspiration in the creation of science that will no doubt determine our chosen solution to this problem. The map of a history of science, as well as its philosophical conclusions, would be predetermined in this way. From the very start of our inquiry, we would be locked into a cycle that we created and from which we would not be able to break free. Perhaps it is so, but the radical scepticism of those who despise history, who believe that history is a mirror that reflects the historian's own spiritual likeness back at him, could only be sustainable if the historian halted his effort at his first contact with texts, immediately to proclaim his final, unalterable and irrevocable judgement. Incidentally, this is what people do who only see the most distant past, just as the recent past, only through the lens of an immediate present proclaimed a perpetual present. But this cannot be so for the historian who considers history as an actual reality. The historian does not prejudge the results of his research when he becomes clearly aware of the difficulties of his work. In the course of his research, he will constantly strive to go deeper so to understand the past better, to grasp with more assurance and more active sympathy the creative thought of the past, into which he will breathe new life and which he will, for a brief moment, bring back to life. What is more, there exists a personal, subjective factor, as is often said, that is impossible to eliminate completely, whether we openly acknowledge it, or we deny it *a priori*. Like all philosophers, all scientists, and all human beings, historians have innate tendencies and particular and unconscious mental ways that are not quite opinions or systems, but that can beget, and indeed do beget, opinions and systems. And more often than not one discusses opinions and systems as if they were isolated, as if they were not born from a human brain, as if they were not directed, imperiously directed, by certain fundamental tendencies. The historian who has a long-standing familiarity with past texts knows this well. And rather than rebelling against the order of things to achieve an imaginary objectivity located outside of the world and outside of science, he tries to find or reconstruct within himself for a brief time the deep motivation that underlay the works that are the subject to the historian's reflection. When the historian studies the work of Cartesian, Newtonian, Stahlian, and Condillacian chemists, who worked and made their science advance in the eighteenth century, he must make himself in turn into a Cartesian, a Newtonian, a Stahlian, and a Condillacian. He thereby understands, or at least attempts to understand, the different mentalities that have suggested our distant predecessors' theories, hypotheses and experimental research. He manages to understand fully, or at least takes the only path that enables him to understand fully, the work of the diverse schools who accepted their research, or of those who opposed it. As for the personal factor, due to the personality of the historian himself, without doubt it will be reduced by this striving for sympathy. Without doubt, it will not be erased completely. What one historian will not have seen, another historian who benefits from the labour of his predecessor will know how to reveal ... But what is the point of insisting? As with all the sciences, history is a science that is made, and that evolves and renews itself as progress in criticism and analysis; it forces us to correct our syntheses, which are never dogmatic or definitive.

Let us turn to things a bit closer to home. In order to reconstruct the thought of past scientists (I am thinking of seventeenth- and eighteenth-century chemists), the historian most of the time has only access to the texts by those scientists published in books, booklets and reports included in the works of various academies, and in scientific journals. The historian can then consult the other texts [in these publications], the review of chemical works in those journals, the writing of philosophers, for example [Pierre-Sylvain] Régis and [George] Berkeley, who paid great attention to chemistry. [He can also consult] studies concerning the practical arts and industries such as metallurgy or glass-making based on chemistry; references to chemistry contained in works that were not about chemistry but likely to reveal the social role of chemistry. To this one can add certain laboratories, or reproductions of the laboratory that one studies in museums such as the Conservatoire des arts et métiers in Paris, commemorative medals, statues, or monuments, drawings or allegories illustrating a number of publications. These things all have their own value. However, I propose that we limit ourselves to texts so as not to be distracted by these parallel discussions. I do not think, although one never knows, that you will disagree when I tell you that, in my opinion, the ultimate goal of the historian of science, as is the case with the historian of literature, is to arrive at a total understanding of the texts he studies. In order to achieve this goal, he can, and must, draw on the serious and fruitful methods that ensured the success of literary history, enabling the literary historian truly to bring the past back to life. I am not going to talk about these methods, and you will not find it hard to agree with me that the historian of science must perform a philological and critical analysis of the texts that he studies.

One must recognise, however, that the scientific work, which is in some ways related to the purely literary work, differs from it due to certain other characteristics that make it difficult to access it. I do not need to remind you that the texts used by the teachers of literature for our courses of *culture générale* often seem to have been written, in a manner of speaking, outside of time and space. In other words, they remain accessible to the intelligent reader of any time and place, without any preparation whatsoever, provided that, of course, this reader knows the author's language. So as not to digress into a discussion on the pre-eminence of ancient or modern languages in forming the mind, I am only referring to French publications. For example, senior school pupils read Racine's tragedies, Molière's comedies, Pascal's *Pensées* and La Fontaine's fables without concerning themselves with the 'exact time' of their publication, since the jewels of humanity's heritage do not carry a stamp from any particular era. I confess straightaway that I am exaggerating for argument's sake, but I am only exaggerating very slightly, and I am referring to the traditional notion of works that are labelled as 'classics'. I know that a large part of this notion is an illusion, and that Taine's criticism, which we will not be discussing today, has long dispelled this illusion; but all the works that I have just named, and many others, can be studied in depth and read in a variety of ways for the reader's education, instruction, and pleasure. One can reflect on what one reads in the context of one's one thoughts, and without being preoccupied with the author, simply to extract intellectual and moral gain from one's reading. One can also read to get to know the author better, to grasp his thought or his system. One can also read for many other reasons; finally, one can combine different ways of reading in order to understand the past better, to understand the present better, or to understand oneself better.

That a first-class scientific work is not exactly the same as a first-class literary work is something that I shall not contest. I shall also not contest the notion that the history of general literature (in the slightly vague sense of literature handbooks) encounters fewer initial difficulties that need to be surmounted than the history of scientific literature. Nevertheless, I do venture to suggest that old scientific publications, even those that present theories that have become definitively and irreparably obsolete, are by no means devoid of educational value and suggestive power. These writings can contribute to the enrichment of the mind as much as those disciplines that we call humanities. But again – and perhaps, you will tell me, just like the humanities themselves – these writings do not immediately deliver their beneficial effect to those who leaf through them hastily or lazily. It is necessary that the historian, by means of an assiduous effort of sympathy demanded by his method, manage to grasp those books of past masters just as the disciples and pupils of those masters would have done. The historian must understand, for example, what science in the making would have been in the eighteenth century. If he manages to understand the mind-set of the scholar whose work he is studying, this work will then become transparent and crystal clear (with the exception of what could be called the material obstruction, to which I shall return). When he reads van Helmont or Lémery, he will imagine, as a background, the social or mental milieus in which these chemists were immersed, but also those to which they were partially reacting. Van Helmont's work, for example, is a product of early seventeenth-century Counter-Reformation. By viciously attacking Aristotle's metaphysics and logic, by attacking astrology, he was acting as a scientist who was ridding chemistry of a number of superstitions, but he was also acting as a Christian fighting against the offensive return of paganism. Van Helmont was inspired by Genesis, and by the intellectual intuition which [for him] God freely bestows on those of his children who know how to pray, toil, and carry out certain moral practices that maintain their mind in a state of receptiveness. The laboratory experiments that he performed over and over again, providing as they do unmediated information about the world, provide only information, and so need to be interpreted, since they do not fully inform. They were combined with prayer and meditation so that the chemist could receive a communication of divine thought into his soul; and this is where true intelligence lies, rather than in logical arguments that force us to repeat ourselves ad infinitum, in other words, without ever advancing natural knowledge. Christian mysticism was therefore combined with the experimental method in order to renew the science of material reactions.

Let us now move on from the first half of the seventeenth century to the second half of the seventeenth century. We come to Lémery, who thinks in a Cartesian style in a Cartesian world. We admire this great man who knew how to inject the clear and lucid spirit of his time into chemical theory once and for all, as he knew how to impose the study of chemical theory on the honest man, who is proud of nothing but aware of everything. He removed from this theory the Paracelsian analogies that had enchanted the joyous vitalism of the Renaissance; [thanks to him] the alchemists' foolish hopes of transmutation of metals became absurd and ruinous superstitions. Generally speaking, he drove out all poetic and allegorical language from science. But we observed a beneficial revolution not only in the theory. The art of performing experiments was modified and filled with integrity. Any laboratory experiment that resembled a conjuring trick, that demanded at least a difficult and secretive sleight of hand, would be considered suspect. Lémery would disregard it. He did not want to be 'secretive in his operations'. It must be possible for everyone to repeat these operations; there was no mystery to his practice, just like his theory, which was first and foremost 'intelligible'.

I am not going to amuse myself with yet more examples. I am also well aware that while sketching too quickly an intellectual and moral background against which the historian of science can bring into relief the properly chemical works, the practical discoveries, and the systemising theories of van Helmont or Lémery, I have not spoken of the definitive contributions that these two great men made to the body of chemical knowledge. I have sought first and foremost to understand the scientific perspective (if we call it so) of scientists who dedicated their lives to perfecting or to creating a new science. I have sought to understand their mentalities and this, I repeat, is of crucial importance to me. From one generation to the next and as science progresses, the scientific perspective changes: a problem that was once central no longer interests researchers, and a minor curiosity that intrigued amateurs becomes a central problem. At the same time as the perspective changes, the mentality is modified: after the cast of mind that was embodied in the effort of systematisation, there soon follows a very different attitude that brings about little by little a new way of understanding. And of course, since science is a collective pursuit, scientists continue the work of their predecessors, even if they transform the theories of their predecessors, either suddenly or by means of a slow evolution. It is not because science, by progressing, sometimes yields to the latest trend and absorbs the same principles of current opinion, that scientific knowledge ceases to be, as Sarton rightly said, cumulative and progressive. And knowledge without doubt does not accumulate like merchandise; it becomes integrated in theoretical monuments thanks to scientists. But history shows that these monuments are constantly being overhauled or rebuilt on new foundations with a mixture of old and new materials.

I am not here today to talk about scientific revolutions or their significance with regard to the progress of the human mind, or with regard to the progress of our knowledge. The historian of science will probably have to present documents, the importance of which would be difficult to overstate, to the philosophers who will attempt to solve the problems highlighted here. After the remarkable works of Brunschvicg and Meyerson, so different from each other, but both basing epistemology on the study of past and current scientific reflections, the historian of science seems not to have any further need for advocates among philosophers to defend his right to exist.

As the importance of the history of science has been established, I shall therefore plead before you only the case for good history of science. I shall not forbid to the historian of science to be himself a philosopher or to support a certain philosophical theory; I shall simply ask him to think as a historian when he does history. While engaged in his work, he should not be concerned with knowing if the conclusions of his work will justify a particular notion of intelligence or human reason. He should not imprison himself in a particular *a priori* blueprint that would indicate to him a ready-made conception of science or of its history. The agreement between the facts that the historian studies and the philosophical theory must happen by itself, without trickery, without prompting of the facts or the theory, without pushing in a certain direction; in short, with no arbitrariness whatsoever. In this way, we may be able usefully to seek to solve the problem that Lalande proposed to us. What are the relationships between *constituent reason* which, in its permanent and essential tendency, is human reason in its entirety, and *constituted reason*, which is the aspect that this reason presents at a particular moment of human development?

Let us return to the history of science; let us suppose that the historian, blessed with good will and insight, has finally succeeded in understanding chemical theory as if he were a pupil of van Helmont, Stahl, Lémery or Macquer. Would he no longer encounter any obstacles, and could he proclaim that he is completely satisfied? Alas, he would still face a difficulty, which has not yet been overcome and which can perhaps only really be partially overcome! Earlier, I spoke to you of 'material obstruction', and I must now talk about it.

I shall make my point once again using the example of the chemist. The latter enters his laboratory to work practically and to observe phenomena brought about by his own work. He may mix certain bodies with others and observe such and such material reactions; he may bring about dissolution, precipitation, evaporation, combustion, calcination, and of course other things that he has or has not anticipated. He records the results of his experiments in writing; he does so, of course, by employing the language of the dominant theory. He explains them if he can by using this theory, or he proclaims them to be unexpected and surprising; perhaps he may be able to account for them by means of auxiliary hypotheses ... Or it may be that, although the texts are precise and clear, the historian is unable to grasp fully their real meaning. Given the reagents brought together, the result of the experiment should, it seems, have been something else entirely. Then a question arises: are the bodies used by the chemist in his laboratory experiments the bodies that we still call by the same name? Often, one should read antimony sulphide for antimony. Other times, one may wonder if the reagents were pure or homogeneous or if they rather contained impurities that modified their properties. The acid in sea salt could contain vitriolic acid, nitrous acid or even muriatic acid [hydrochloric acid]. Other times, one does not know and one has no way of knowing. When the historian therefore wants to make himself the contemporary of the authors whose work he studies, he will indeed be ill at ease. He will understand the words, but the laboratory floor will appear to give way under foot; the past scientist will jealously guard his secret. And as one must always learn from one's failures, the historian will notice two things that might have eluded him. First, he will realise that the advances in chemistry are even greater and more important than he could ever have thought at first. Since pure bodies, placed in well-corked bottles and bearing the current chemical formula on the label, were not kindly offered by nature, these reagents are, one could say, almost without exaggeration, created by theory. At least, without the use of the theory that directed chemists' research, science would never have obtained and studied them. Second, he will realise that past scientists did not record the results of their investigations in order to facilitate the work of future historians of science. Chemists, like physicians and naturalists, do their utmost to solve a universal or particular problem that attracts their attention. As they address their contemporaries, they neglect to insist on things that are not up for discussion and that have received scientists' unanimous and spontaneous approval. What good would it do to expand on the foundations of the theory, which no one would think to discuss, because everyone takes them for granted? These neglected things have become of primary importance for the person who studies [those texts] several centuries after the death of their authors, and the historian should first and foremost specify them in order to grasp the orientation of a past mentality. In order to restore them he should, in a manner of speaking, dig below the surface of the theory. All this does not make reconstructing the past an easy task. And it is for this reason alone that the history of literature and the history of philosophy are infinitely more advanced than the history of science. Certainly, the historian of science is not devoid of tools. In the eighteenth century, for example, there were violent polemics, which sometimes degenerated into passionate arguments between Cartesian chemists, Newtonian chemists, Stahlian chemists, Condillacian chemists and the chemists who were claiming complete experimental empiricism. Those polemics and disputes provide us with precious insight into the scientists' state of mind. But even there one must advise prudence. The historian's critical mind must intervene in the interpretation of texts, which he cannot be content just to collect and gather together. The historian who wants to understand the real thought of scientists cannot simply examine the arguments put forward by one or the other of them. This is because theories are established and maintained because of the satisfaction that they give to human intelligence, whereas they are attacked and defended on their lines of least resistance. Finally, the data relating the problems that scientists strive to solve are not fixed; in fact, they are modified as one discovers new facts or as one imagines partial answers to the questions raised. There is a sort of gap from one generation to the next, which causes variations both in the orientation of mentality and in theoretical and experimental data. It is in this way that for the Stahlian chemists, who had understood that the calcination of metals was combustion, but who believed that by burning, bodies would release their phlogiston, the increase in weight of metals became an irritating or amusing enigma that had to be explained on the margins of the theory. But Lavoisier discovered that the products of the combustion of phosphorous and sulphur weighed more than the original phosphorous or sulphur. The calcination of metals was therefore no longer an exception. By explaining why the lime of tin or lead had a greater material mass than lead or tin, he no longer explains an astonishing and isolated fact; rather, he accounts for all phenomena of combustion at the same time.

Allow me to carry on no more. I must observe again, however, that I have spoken only of the preliminary work of the historian here, as there is still another work that is just as important. Once the historian has succeeded in making himself the contemporary of the scholars whose theories he wants to expound, his work is only half done. He must write his work in such a way that his reader is also able to penetrate into past theories, but without any, or at least little, effort. How can one achieve this objective? I cannot tell you today, and moreover it is up to the reader to decide if the historian has succeeded in resurrecting the past.

## Note

1 Talk delivered at the Unit for the history of science of the Centre international de synthèse on 18 January 1933.

# Appendix 2: The tribunal of history and the theory of scientific knowledge[1]

Will you consider today an issue that has very often occupied my mind in the course of my studies? Will you be so kind as not to dismiss this issue *a priori* and with disdain, even if upon its utterance you consider it strange, metaphysical or unexpected? I will warn you straightaway that the question that I am putting forward is outside the field of facts that the historian of science endeavours so laboriously to interpret. However, when the historian of scientific thought hears the compelling calling of his vocation for the first time, he urgently asks himself this question. When he stops working for a moment to examine his philosophical conscience, he asks himself this question again. When his work is complete, and he wants to take an overall view of the results obtained from his formidable and sustained effort, he asks himself this question once more. You see that it is important, and, without further ado, I will risk delivering it to your eager attention, so here it is: 'Does the tribunal of history have the power to deliver judgements that are able to close the debates raised by philosophers who are supporters of different theories of scientific knowledge? Can it, as a last resort, pass a verdict that condemns some theories and approves another theory submitted for its review? If it were possible to answer 'yes' to this question, if the precise fact, the positive fact, the objective fact (to employ a pseudo-scientific jargon that is alas still too widely employed), if even a combination of these facts chosen and presented with skill were to impose a peremptory and definitive opinion on the philosopher, if an equally intangible and absolute decree could halt debate between different thinkers seeking to understand the workings of the human mind, the tribunal of history would have rendered any speculative research unnecessary. The theory of scientific knowledge would no longer belong to the field of philosophy. Without doubt, philosophy would not be annihilated, as it would be able to use the information supplied by the scientist to solve another given other problem over which science currently has no grip! But its importance and its field of research would be reduced significantly; it would have suffered a defeat. Thus, history in an instant would have vanquished the reflection on science, the argumentation made with the aid of science, and the dialectics of scientific discussion. The philosopher would have provided humankind with the theory that won approval from the tribunal and, following this trial, he would probably be carried away by his victory: he would have solved for all eternity an enigma that had been put to him, and that henceforth will be put to him no longer! He will lose, perhaps for a brief moment or forever, critical reasoning and modesty; but let us just leave him to his delight!

Let us observe the crowd of dissatisfied people who, following the established and in fact inaccurate formula, have refused to burn what they had loved, as in the end those dissatisfied people did not love for no reason; otherwise, they did not love at all. Their former conviction was based on the spontaneous and reasoned judgement that they formed based on their studies and extensive reflections. This conviction could not, as is the case with a new faith, rely on an authoritarian decree which by itself has the force of law and against which nothing shall prevail. They clearly knew that to err is human, that the previously professed theory is continuously subject to improvement and revision. Why would that not apply to that which official philosophy supported by official history wants to impose on them by force, avoiding all debate? Our disgruntled philosophers will restart the proceedings: they will consult the archives, they will weigh up the resulting evidence, as it were, and will examine the documents that constitute the incriminating evidence ... And then, perhaps, they will admit that they were mistaken, that history as it has been presented to them has freed them from an error. If they discover why they made a mistake, peace will return to their souls. They will accept the theory imposed by force on them, without the assent of their minds, but they will accept it freely, as their amended judgement reveals the excellence of this theory to them. It could also be the case that, despite their good will, our thinkers have to refuse to accept the verdict of the 'tribunal of history'. Turning against the prestigious philosopher and historian who would like to impose their points of view by any means possible, they will continue to invoke the spontaneous clarity of their mind. They will declare with fierce energy that there is no judgement against judgement, that force and reason are two heterogeneous powers in conflict with each other. They will wistfully add, taking inspiration from the title of Jules de Gaultier's work, that official philosophy is something quite other than philosophy!

Well then, as we have discussed hypothetically, science and philosophy cannot put up with authoritarian despotism of any sort, they refuse all regulation and they would rather temporarily vanish than allow themselves to be constrained. It then follows that we must recognise that if the tribunal of history cannot equip with weapons the secular arm to sacrifice or silence those who do not comply, if it only has persuasive force to impose the truth, then its verdict cannot be considered immune from revision. Regardless of the intellectual or moral authority of the group pronouncing the verdict, this verdict does not draw its power from what it is, but from the rational agreement of those who accepted it! We can therefore conclude that if the philosophical history of science has several investigating judges and even several barristers, it could not, fortunately for the philosophers and for the future of human thought, constitute a powerful jury that would decide what the truth is based on a majority vote.

But should we have to swing from one extreme to another? As we have been forced to answer 'no' to the question asked earlier, and we have recognised that history cannot use its own authority to decide how the human mind really works, should we abandon the history of scientific thought?

Without doubt, some philosophers, and it is with regret that I cite the Cartesians among them, have denied that history, even history of thought, is of any real interest. They have said that we can learn nothing from old science, from the science that has been completely overtaken by the awakening of thought, the science that is irreversibly lapsed. Why should we trouble ourselves with scholarship that is just as unnecessary as it is vain, by studying it instead of letting it be forgotten?

Contrary to those who claim that all research is eternally current, which is the same as to say that 'the wind bloweth where it listeth', and who declare for this very reason that the history of scientific thought has no value whatsoever as knowledge, we must mention those who believe that scientific evolution is without interest because it [only] adds a few details to their determinism, which they assume without further examination as a consequence of their *a priori* conception of human evolution. In an admirable talk that I had the pleasure of hearing in Coimbra during the recent International Congress of the History of Science, Arnold Reymond drew our attention to the sociological theory of the history of science. The Marxism taught in Moscow, for example, turned science before the advent of the Soviets into a product of the bourgeois society that reflected this society much more than it enables us to understand the world around us. The 'tribunal of history' is probably, in this case, obliged to give a verdict, but it does not do so freely. It will have to be content with demonstrating by accumulating facts, that the determinism that has been established *a priori* fits with the succession of events. A clever history of these facts will put this conclusion beyond doubt!

There are other examples. One of them is August Comte, whose mind was not built for the patient and meticulous study of history. He accepted as intangible dogma what he called the law of three stages – specifying that always and everywhere the theological mind of the earliest stages gave way to the metaphysical mind, which in turn was finally replaced by the progress of the ultimately victorious positive spirit. However, Comte wanted the history of science to add the support of the authority of facts to its theory. How could it be otherwise, since the truth that he held had been conclusively established? Comte was the first to suggest a Chair in history of science at the Collège de France. He understood that the new professor would be a benevolent judge or an eloquent advocate for the new philosophy. By discovering the value of the history of science and showing it to the world, the victorious positivism had hoped to enslave it once and for all.

Should the history of science be reduced to a pedantic and worthless heap of random and surprising facts that would entertain an old scientist on a Sunday afternoon? Or should it be made to be a foot soldier for a philosophical theory concerning the physical world, the structure of the human mind or society, thereby sacrificing its own nature and autonomous effort due to this zealous and servile attitude?

Between these two horns of the dilemma cleverly put forward by a learned and confident dialectic, we shall attempt to discover the royal road or the harsh path that will enable the history of science to show that, although it cannot hope to replace philosophy, and although it cannot abandon the *a priori* entirely without becoming inconsistent and vague, it can at least illuminate the meditation of the philosopher who elaborates a theory of knowledge. History of science can also guide the attention of the psychologist, the sociologist and the scientist who will perhaps make important discoveries in their different fields thanks to their contact with it.

Today is not about me turning into an advocate for the history of science that has not yet fully triumphed, and which in the scientific world reckons with many who regard it as futile or pointless. We shall not linger to show that past science, lapsed science, the science that is definitively and practically unusable, can offer joys to those able to appreciate it, similar to the joys that art provides to its loyal students. We shall not demonstrate here, that, besides their 'spectacular brilliance', both Renaissance science and art (to give a specific example once again) reveal in a similar way that, at a precise moment in history, human intelligence and sensitivity endeavoured to penetrate nature, with which they aspired to be in perfect communion. Finally, this is not the place to investigate how those who unearth documents or collect texts (who in other areas of human activity are universally revered as soon as their discoveries reveal something original about humanity's past) are sometimes despised by the scientist or the populariser of science as soon as they wander into their domain. We shall not say, as this specialist or populariser would, that the accumulation of facts or historical details only gives satisfaction to idle and misguided curiosity that was unable to be of better use elsewhere. We will accept and admire even more the patient research of conscientious scholars as their discoveries support our effort. Our discipline is not within the competence of the philosopher, who gratefully accepts the kind gift personally delivered to him.

It is true that the philosopher takes an interest in past science and in the past scholar's thought not only because of a passionate love 'for something we will only see once', nor just for its immediate emotional value, in fact not at all for its emotional or aesthetic value. Our philosopher does not lack sympathy towards past generations, although he perfectly understands the ties of solidarity that bind together in some sort of spiritual fraternity the men who, across all times and in all places, are devoted to the search for truth. Moreover, his effort to recreate the mentality of his distant predecessors in his soul is sincere. All the same, he knows that the lapsed theories that he is resurrecting for a short time and by which he is captivated, teach us nothing about the world. These theories are now only worthwhile as tools or instruments that enable us to understand the human mind better; to ascertain the value of human knowledge (if this is possible); to work out why constantly progressing knowledge exerts such a magnetic attraction on the soul of each of us; and finally, perhaps because of all this, to render the most noteworthy services to future science.

Why, all of a sudden, has the thinker addressed the history of scientific thought, expecting it to solve the fundamental questions that up to now seemed to belong exclusively to the domains of dialectics, analytical criticism or general metaphysics? Why has he appealed to the direct authority of the 'facts' that are supposed to establish, solely by means of being brought together, the 'true theory of scientific knowledge'? This is the problem that we shall address here.

I do not need to demonstrate that the history of science was not completely neglected before philosophers decided to use it as an argument – and sometimes even a sledgehammer argument – to introduce their theory of scientific knowledge. That the history of science was practised, however, on the fringes of serious study – under the influence of many different and sometimes even opposing preferences and motives on which I shall not dwell – is something that becomes clear to someone who studies the writings dedicated to it in former times. History of science up to now was only of interest to researchers on holiday, scholars of great value, and the masses introduced to the triumphant saga of civilisation by enthusiastic popularisers. But, I tell you, the day has come when the history of science has been invoked by scientists themselves, in order to prove that their way of conceiving of the aim, object and structure of physical theory is indeed correct. The positivist researchers in thermodynamics, under the direction of their leaders Mach, Ostwald, and Duhem, relied on the works by researchers of all times and places, in order to establish that mechanism and atomism, regarded as the core of scientific theories since the seventeenth-century 'renewal', did not play, despite appearances, the prominent role that most philosophers attribute to them still nowadays. For them, it is an illusion to believe that atomism and mechanicism would act as a guide for the physicist. They were and still are only imaginative by-products, perhaps as noisy as the famous fly in [La Fontaine's] *The coach and the fly*; however, we have to eliminate them as soon as possible so that active intelligence, finally rid of its useless parasites, can work at full capacity without wasting any effort.

You all know that the work so cheerfully undertaken has not been fully accomplished, that it did not have the scope that its makers had attributed to it, that it did not appear to be entirely convincing, and that it did not silence and convince the atomists or the supporters of the mechanistic method of their error. I would remind you here that Rey, in a well-known work that Duhem has discussed at length, endeavoured to demonstrate that, far from being sterile, 'figurative hypotheses' at the very least produced fruitful inventions and discoveries. Since I do not have to go into the current controversies in the theory of knowledge, I will make do with saying again that Brunschvicg's and Meyerson's works, which are so important, so admirable and so different, relied almost constantly on past science in order better to penetrate the human soul.

That the history of science bestows the most noteworthy services on the theory of knowledge, that it aspires to be part of *culture générale* or, to evoke Sarton's words, that it forms the basis of the 'new scientific humanism', is a far-reaching consequence of the thermodynamicists' attempts that we have just discussed. Had they done only that, and all their philosophy had collapsed (you see that I envisage the worst-case scenario), still their contribution to science, to the theory of knowledge and the history of science would have been praiseworthy.

Some slightly naïve historians thus could have believed that the chronological accumulation of scientific texts would reveal the true progression of our intelligence directly by itself, excluding all interpretation, criticism or commentary. They asked this overwhelming body of texts to reveal the truth by silencing once and for all philosophers, who were invited to stop taking up public space with the deafening clamours of their pointless disputes. With this objective achieved, [the historians] offered unlimited thanks to history that had delivered all its secrets to them. Just listen to Delacre artlessly expressing his joy: 'The history of chemistry since Black', he writes, 'was made, in a manner of speaking, in advance. There is one strange thing, namely that nobody had written it down. The most demanding historian will find nothing to complain about, his role has been completely mapped out ...' 'I have spent too much of my own life', Delacre adds, 'amid theories, I had to teach them too much not to have developed an irremediable aversion to them. I suffered this aversion without being able to justify it. Today, with my book finished, I am completely satisfied since my scepticism was taught by history'. The theorists have been vanquished; empiricism, thanks to history has achieved total victory. The experience that no thought will fertilise, criticise, or inspire will reign supreme. Daring speculations have been relegated once and for all to the bookshop for worthless novels.

We must add that Delacre, according to his formal declaration, was already an empiricist before interrogating history. If he had been a mechanicist, an atomist, or a theorist, the response that he would have received [from his sources] would have perhaps been something else entirely. Let us remark at least that it is as a result of a prior value judgement that he dismissed Mendeleev's table [of elements], and that he appended to Wurtz's name the seemingly offensive epithet of 'professor of 1860'.

Faced with these and other similar cases, upon which it would not be wise to dwell for too long, the philosopher leaves the historical work that he has begun for a moment in order to hesitate in a melancholic way. 'Is history nothing more than a mirror reflecting back to the thinker his own spiritual image, just as dreams, following Freud, are nothing more than gratifying manifestations of our deepest and most hidden desires?'

But let us not fall into a sterile and mournful despair. If we have doubted for a moment the value of history as knowledge, this could be because we have abruptly stopped our inquiry at the very beginning, by asking a 'yes or no' question and therefore refusing all rectifications and modifications. Similarly, the pragmatists, by considering science from the intangible point of view of absolute subjectivism, have doubted the cognitive value of all theoretical science.

Does this reconciliation between some aspects of historical and scientific knowledge not naturally suggest to us, however, that if history must be considered to be a science, if it must be studied as a science, it follows that it will be impossible for history to be placed outside of science, in order to judge science? Therefore, with the exception of the questions that all historians must ask themselves concerning the accuracy of recounted facts, their chronological accuracy, the authenticity of documents, and the wise interpretation of texts, they will of course freely accept to submit their conclusions, which they know are provisional and perfectible, to a philosophical criticism that nothing will supplant.

But I am not going to make myself look absurd in front of the director of the 'Bibliothèque de synthèse historique' [Henri Berr] by trying to demonstrate that history is a real science, that its method is only technically different from that of all the other sciences; because all of this follows irrefutably from the work of Berr himself and his eminent collaborators.

We should note at this point that maybe all theories of scientific knowledge with which philosophers have provided us find a new field of application in the practice of the history of scientific thought. Discussions, far from being closed by an arbitrary decree, will continue at an even faster pace as the history of science is only starting to attract the attention of the academic world. This discipline is not yet well advanced, as it is still directly dependent on the motives that have guided researchers to study it with fervour, in short, as it is only just out of its long very first infancy. One should not under any pretext try to close down these animated arguments that stir so many ideas, and which are a first-class stimulant for the mind at work. That said, I shall not hesitate to take up my responsibilities by submitting to you a few personal opinions, which I ask you to criticise rigorously and indeed mercilessly.

We should note first and foremost that the theory of knowledge, the reflection on the value of knowledge, and the criticism of scientific hypotheses, emerged well before philosophers thought of studying lapsed science, taking an interest in long-forgotten past theories, or evaluating the evolution of science throughout the ages.

Not only has the theory of knowledge preceded the history of scientific thought, but it has also guided the latter's first steps in the field of philosophy. The theory of knowledge has guided its attention, has proposed to [the history of scientific thought] a series of questions to answer; it has kindly asked it to provide an abundance of examples to be used either as evidence or as illustration of its claims. We should not complain about this way of proceeding. We believe that if the historian had no preconceived ideas when approaching his work, this work would turn out dull! Note that if in your opinion, the experimental method can and must be used in the history of scientific thought, this experimental method is genuinely far from being the same as empiricism in the strictest sense. Collecting documents is not history, and collecting measurements is not science, even if history cannot do without documents and science without measurements. History is nothing without the historian's intelligence, and physics is nothing without the physicist's intelligence. We should assert once more and without fearing the contradiction of the supporters of pure experience, that the facts that history unearths cannot reveal the path followed by the human mind by their mere assemblage.

So, you ask, how can we come to catch a glimpse of the route of scientific progress? Note first that as modest as the historian's mind may be, it cannot remain totally inactive. By making a wise selection among the innumerable documents that attracts his attention, he already imparts a direction to his work, but these chosen documents do not assemble themselves and remain heterogeneous.

The historian is often required to reflect for a long time in order to grasp the true meaning of the texts that he has gathered and organised. He will examine them in every possible way in order to be able to extract from them the riches that evaded him at first glance; he will critically analyse them and interpret them in various ways, he will come up with a number of heterogeneous hypotheses about them, and at that point, he will not be afraid to introduce an *a priori* derived from his personal opinions and reflections. However, his task is far from complete: he must not be content with selecting and understanding documents. He still, following Enriques' very appropriate and precise formula, will have to build the historical document that he will deliver to his reader when he reaches the conclusion of his sustained effort.

Without doubt, such a way of proceeding will then prevent the historian who is aware of his own method from claiming absolute objectivity, which would for ever assure the certainty of his conclusions. The history of scientific thought cannot demand philosophical approval. From this point of view, and since history, like any science, can progress, we can clearly see that the tribunal of history does not have absolute competence and that it is entirely unqualified to pass a verdict without appeal.

However, the historian's meditations will enable him to shed light on those philosophical problems that he would have liked to solve. Perhaps these meditations make him see an aspect of these problems that he would not have seen without them.

In order to conclude this talk, which I could extend for a long time without fully covering the vast topic at hand, I am going to invoke my modest experience to show you how the history of science can project a glimmer of light on the complexity of science.

Let us consider a specific example; when I began my work on the history of chemistry, I read a celebrated paper in which Robert Boyle claimed that, having exposed several bodies such as bricks and certain metals directly to an intense flame, the weight of these bodies increased. He explained this fact by claiming that the substance of fire itself gets incorporated into the matter on which he was conducting the experiment. I have thought along with other commentators that here is a theory of combustion of certain bodies that is truly bizarre. Why did Boyle choose it without examining other theories? Why did he not assume, like Jean Rey or Du Clos, that the increase in mass observed on the scales was due to the penetration of air or impurities present in the air into the body in question? The chronological succession of facts teaches us that Boyle's theory experienced lasting success and that it was nevertheless immediately contested by P. Cherubin, but it does not enable us to grasp the psychology of the great English scientist.

Indeed! By later studying the same text in different circumstances, and with regard to theories of light emission, I suddenly understood that Boyle's interpretation, far from being a direct consequence of his research, was actually the result of the way in which he had posed his question.

Note the contrast between the minds of the various researchers who were trying hard to solve similar experimental problems. Jean Rey, after conducting experiments, asked the question of why lead and tin increased in mass when charred. And the response he gave would have remained within the eternally debated questions if a century later chemists had not decided to capture and weigh aeriform substances or gas. Robert Boyle gave his work the title: *New Experiments to Make Fire and Flame Stable and Ponderable*. It is clear that such a programme influences the interpretation of the experiment.

Note that Robert Boyle, who might not have known what the results of his experiments would be at the exact moment when he attempted them, could not contemplate asking why some bodies increased in mass at the precise moment that they were brought into contact with a strong burning flame. In fact, he wanted to examine the facts directly, he wanted them to resolve the dispute that had arisen among many philosophers, concerning the nature of light. He wanted to know whether light was a substance or an accident. As the quantity of ponderable matter grew over the course of a laboratory experiment, and as this increased mass could not, according to him, be attributed to the intrusion of a volatile body matter fixed for a brief time into another solid body matter, he concluded that in all probability this volatile matter is identical to light, and that light is a body. Gassendi was therefore right to say that light is a substance; from then on, light could be considered one of chemistry's reagents.

It would take too much time to pursue the question, which I will leave to your reflections, regarding the lack of fit between the question posed to nature and the answer that nature itself gives. In the end, if we do not enter Boyle's state of mind, we cannot see a direct and necessary link between the problem formulated as 'is light a substance or an accident?' and the conclusion of the experiment, namely 'the weight of some substances increases when they are subjected to the action of an intense and prolonged fire'. Perhaps it is in this lack of fit between theory and experiment that lies the doubt that affects the whole of physical theory ever formulated. Perhaps it is in this lack of matching between facts and physical theories that the theory of knowledge will manage to discover why no notion can be given an unalterable definition for eternity; why no theory is truly protected from a possible reformulation; why science evolves slowly when the scientists' mentality changes equally slowly under the pressure of diverse causes; why the same science undergoes a sudden revolution when, due to the discovery of a new and fertile point of view, the scientists' mentality suddenly changes.

All in all, it seems clear that the study of the history of science would eventually heal the philosopher (if the disease could be cured) of the strange habit of wanting to present definitive *a priori* or *a posteriori* concepts, on which the mind could quench its thirst for certainty, and which one could rightly call concepts of divine right.

If then the history of science (or a reflection on certain episodes in the history of scientific thought, if you prefer) teaches us quickly to gather together all the possible experiments that a hypothesis could inspire; if it teaches us quickly to discover all the possible hypotheses stemming from the same experiment; if by this it succeeds in giving our mind, constantly kept alert, and clear of all dogmatism as well as of useless and sterile scepticism, a little of that active plasticity that is a condition of all fruitful research, then it would render a service both to science in the making, and to the soul of the researcher, liberated from inconspicuous and tiresome routine. The value of this service would be impossible to overestimate.

Would this not be better than attempting to assume the role of a board of examiners or a criminal court, which awards high marks to some scholars and low marks to others, or which condemns some theories and accepts others, without being able to impose a sentence?

I have given you my opinion tonight about just a few of several questions that may be controversial. I have spared you some others; I have spoken enough. I urge you all to tell me what you think of the views that I have just presented to you, probably a little too schematically.

## Note

1 Talk delivered at the Unit for the history of science of the Centre international de synthèse on 23 January 1935.

# Appendix 3: The *a priori* in scientific theory and in the history of science[1]

With a boldness that frightens me, I will today attempt to focus your scholarly reflection on a problem so vast that a huge tome filled with closely packed writing would not cover it sufficiently. I propose to investigate whether it is true that the *a priori* necessarily plays a crucial role in the formulation of all scientific theory; if this is true, I propose to investigate what its role is.

If I thus dare to ask you to examine a series of questions, it is because these questions arise almost constantly in the course of the historian's numerous investigations. It is not our responsibility to solve them entirely, but we can hope to shine some light on them.

I will still ask you to examine these questions in the actual form that they take in the mind of the researcher who painstakingly works to reconstruct the development of the past scholar's thoughts. Then I will, and this is my final request, ask you to put to one side for the time being the different ways of conceiving the *a priori* that the greatest thinkers, of whom humanity can be proud, have put forward over the course of their memorable research. We shall not talk about the disagreement over innate ideas, the conflict between rationalism and empiricism, the endlessly resurgent dispute over realism and idealism, Kantian critique, evolutionist hypotheses, causality, time or space; and, so as to end this overly long list, we shall not talk about the many theories of scientific knowledge.

Let me reassure you: my aim is not to sweep away the scattered or systematic theories of our close or distant predecessors. I will not be so absurd as to ask you to ignore or rather to pretend to ignore philosophy. In fact, I do not need to tell you that the primary aim and the greatest reward for the history of scientific thought is to be of service to philosophy … However, the historian of science must be constantly on his guard to resist the seductive and highly dangerous temptation that may otherwise steer his effort off course or render it ineffective and sterile. If before starting his research he indulges in discussions of the various prevailing opinions on the nature of human knowledge and intelligence, as he really would like to do, he would run the real risk of never being able to commence his work. I dare say that this would be a shame and that the philosophers themselves would regret it.

Having made the formal pledge not to get lost down the rabbit hole of preliminary questions and nuanced criticisms, let us straightaway bring the historian of science face to face with little known old texts that he will endeavour to grasp. In order to focus our thoughts, let us assume that these texts relate to chemistry. We then observe that our historian will feel disorientated as soon as he begins his reading; the obsolete language of the texts' author probably will not seem absolutely alien to him: without doubt he will understand most of the words employed. However, two things will remain completely obscure. First, the historian will often feel at a loss to identify the bodies and reactions that the past scholar has described quite clearly in order not to leave any doubt in the mind of his contemporaries working in a laboratory that we are not permitted to enter. How can one attempt to determine the true composition of the ingredients used? How can the results obtained be determined? Let us also not forget that, on the one hand, before the creation of modern chemistry and chemical analysis, reagents could not be defined with all the precision that seems natural to us today. On the other hand, in an era when the experimental consciousness was still obscure, technical procedures consistently presented as mandatory, and understood as infallible, might have been proven correct by the diligent researcher, but they might also have been just a development or a justification of the author's theory. How can one attempt to distinguish between theoretical dream and observed fact? Did our distant predecessors even ask themselves this question? In this case, why superimpose the demands of our current mentality that we are used to considering obvious, onto a past mentality that is significantly different from our own and that would presumably not know of these demands?

Secondly, sometimes the very chain of claims and arguments of the author under study seems at first sight to be completely unintelligible today. After feeling briefly stunned, perplexed and undecided, will the historian of science declare that a work by a Paracelsian or by a sixteenth-century alchemist, in which he can only see a pointless display of verbalism, is absurd, devoid of meaning and without value? Beware of the attitude that you take. This is because if, for the sake of objectivity, which many mistake for wisdom, you persist in only investigating the direct links between the observed facts and the proposed theories, if you only want to employ the rules of positive and prudent logic in order to evaluate a text, if, for fear of subjectivity, you do not put your very soul into your working method, you will in fact declare that the thoughts of our distant predecessors are just a sweet folly, of which neither the reasons nor the motives can be understood. Consequently, you would render the historians of science's labour pointless, at least with regard to a not-too-distant past that you will declare impenetrable.

Let us not employ the method to which I have just alluded. This is as irreproachable in theory as it is sterile in fact, and I much prefer to work with concern [about the exactness of my interpretation] than to accept immediately hopeless conclusions ... I told you a moment ago: we will not give in to the temptation of immediately studying all the theories concerning the *a priori* because life is short, and the demands of our labour are immense. I add this now because we are in the presence of a danger no less grave, and of a temptation of positivity that once again might appear unquestionable. Do not be afraid to make use of all the resources that your creative imagination and inventive intelligence graciously offer you in order to seek out a thought that is impenetrable to our scholarly methods of analysis and synthesis. Do not be afraid to forge ahead boldly or daringly, do not be afraid to employ all your active sympathy to seek out a theory of which the text gives you the outward appearance but not the soul. In short, do not allow your morbid fear of being duped by your spontaneity to kill or paralyse your spontaneity ... And certainly, you will end up making mistakes in interpreting texts; certainly, you will end up making mistakes. But who will stop you from checking your suppositions and testing your hypotheses? Who will prevent you from criticising your work yourself by applying the conclusions of your reflections on other texts? If you have got it right, other texts from the same era – it does not matter whether by the same scholar or by other scholars – will suddenly become clear and transparent ... Remember that, with regard to antiquity, Mr Enriques told us here how he was able to reconstruct the thought of Parmenides, which appeared impenetrable to all ... And then is making a mistake such a terrible thing? Although to err is human, and although the historian of science is prone to error, for him or any other human being only to persist in committing such errors is of the devil.

The history of scientific thought therefore permits at its core an *a priori* that is something completely different from a crude working hypothesis that one can either use or leave on the shelf like a carpentry or laboratory tool. The historian of science first assumes along with Terence and Mr Sarton that nothing human is alien to him. The historian of science assumes that he can and must – by means of an effort which I will not take upon myself to explain psychologically – penetrate the mentalities of the most diverse thinkers who aimed to understand the world. Finally, the historian of science knows that it is in his soul that he must succeed in bringing back to life, or at least in reconstructing, the state of mind of the scholars whose doubts, disappointments, and triumphs he describes. The historian of science knows well that his personal faculties are perhaps not up to the immense labour that the task he has undertaken requires of him. He knows that his intelligence, the sympathy he desires to have, and his historical imagination are very limited, subject to certain aversions and strange weariness ... But he also knows that he does not have any 'reagents' available other than himself; and so, because he realises that for us humans, thought, far from being something that floats around, is always the thought of a thinker; because he knows that he does not want to engage in the vain and pointless game of studying thought as if it were a thing, an isolated and inert object made of words, he rather places himself with timid boldness at the very centre of his effort, like a spider in the centre of its web. He offers his conclusions humbly, he does not seek to impose them; he asks other historians to check or correct his own assertions ... He finally proclaims that a method that is in part *a priori* must be used without a moment's hesitation in order to study the *a priori*.

We have now arrived at the point when I have to reveal fully to you the thesis that will be the object of your fruitful discussion. First, and in virtue of my working method, I shall be obliged to expand on the meaning of '*a priori*' slightly beyond the definition provided by Mr Lalande's admirable *Vocabulary*, which I shall quote to you first of all. One calls *a priori* notions that are independent of experience, 'at least relatively speaking, that is to say that experience presupposes them, and does not suffice to explain them, even when they only apply to experience'. Mr Lalande further adds: '*A priori* does not indicate chronological precedence but rather logical precedoence'.

Now, since we do not wish to meddle with this excellent definition, we shall instead have to adapt it to the way of seeing of the historian of science who, over the course of his work, has acquired the conviction that certain characteristics of theories derive as much from experience and observation as they do from the researcher's mentality. If one acknowledges this, the *a priori* will not just represent all the concepts in place prior to experience and on which experience relies. The *a priori* will also represent the fundamental tendencies that produce these concepts; or to put it in scholastic language: to the *a priori* in actuality, which is made a reality by the first notions, we shall add the underlying *a priori* in potentiality which, on contact with the experience of life (and not only scientific experience), effectively becomes *a priori* in actuality.

And now, we can affirm that the *a priori* is not and cannot always be similar to itself everywhere; or rather, there is not only one *a priori*, but several different *a prioris* that are sometimes heterogeneous and incompatible.

This declaration, which may astonish you, should be backed up by an irreproachable demonstration and presentation of several pieces of evidence, because I know all too well that it would not suffice to affirm dogmatically; above all it is necessary to demonstrate. Nevertheless, so as not to abuse your patience, for which I am grateful, I shall only show you the most fundamental types of *a priori* that my strolls through the history of chemical theories have enabled me to gather and identify.

Let us begin by briefly studying the ways of seeing that captivated sixteenth-century chemists and doctors: the *a prioris* of expansive thought. By *expansive* thought, I mean thought which rushes forward tumultuously and simultaneously in all the directions through which it can fight its way, which advances constantly and irregularly without stopping to consider the ground covered, and without attempting to build a theoretical monument! By giving themselves over to the impulses of their soul, and by accepting to model the development of their theory on the very rhythm of the mind's instinctive life, these enthusiastic researchers were then on the lookout for extraordinary phenomena that they could admire with pleasure. These researchers, who, because of the disposition of their mentality, did not consider protecting themselves from errors or superstition, believed with naïve good faith not just everything their colleagues said, but also everything beautiful and unexpected that they imagined. Laboratory experiments, as well as common experiences, were only a starting point that set off their confused need for artistic symbolism. Doubtlessly, sixteenth-century medical chemistry, which was inspired by laboratory work combined with an abundance of analogies and correspondences, made some manifest progress in the art of healing and making remedies that it would be pointless to contest.

Now listen to a few of Croll's assertions from his long-celebrated treatise, *Signatures of Internal Things:*

> 'Just as earthly flowers show us the colour of the stars, when the meadows are blossoming, so the stars show us a celestial meadow when it comes to the flowers that represent us.'

> 'There are as many kinds of colic in men as there are forms of wind in the world.'

> 'There are as many types of bones in the human body as there are species of wood in the world.'

> 'All walnuts ... have the signature of the head ... the hard shell resembles the skull. – The skin that encloses the meat represents the meninges or membrane of the brain. – The meat absolutely resembles the brain, because of this it chases poisons out, and crushed with the spirit of wine, greatly comforts it, provided one applies it to it as a poultice or plaster.'

You want to stop a bleed? Know that 'the decoction of red sandalwood with wine stops the incontinent flow of blood. – The hematite stone and corals, when placed and enclosed in the hand, again, stop [the flow of] blood. – The sixth class of geranium, which has a red root, is also admirable for stopping the flow of blood. – The male anagallis, which is of the colour of blood, pressed into the hand until it is hot, stops the blood, even when the vein has been cut.'

Without lingering to admire the picturesqueness and the incoherence of the triumphant thought of the sixteenth century, we should remark with Fontenelle that these quotes, to which I could have added almost *ad infinitum,* are pleasant to the imagination but unbearable to reason ... unbearable to reason indeed! Impossible to perceive the logical process made by the mind of the scholar; impossible to believe even for a moment that these extraordinary remedies had been dictated and tested by experimental method; impossible even not to be repelled by the apparent absurdity of the statements. But why then are these statements almost always pleasant to the imagination? Let us attempt to solve this puzzle. For those who sense the charm of sixteenth-century expansive thought, with which we have been in contact for a moment, does it [expansive thought] not immediately invigorate this non-abstract and almost infinite need for generalisation and sympathy that today is dulled, but which characterised the spawning of our intelligence from our early childhood onwards? Could we not, by playing around to some extent, find countless variations and embellishments on the themes that come to illustrate the words you have just heard ... But, I hear you ask, where is the *a priori* to be found in this expansive thought? An insightful question, to which I shall respond first of all that the *a priori* does not show itself as prior notions with regard to a mind that perpetually moves forward without ever reconsidering its conclusions, which is furthermore not concerned with coherence and that does not perceive illogicality. It remains entirely caught up in the very process of the development of thought. Next, I shall respond that if we could reveal these processes, this would give you the very means for understanding or reconstituting the state of mind that avoids any critical analysis. And yet I believe these processes can be split into two groups: the first is an analogical inference[2] which, from the point of view of reflective thought, may seem hypothetical and virtual, but which the soul at work believes to be real or one could say material. The second consists of activating this analogy, that is action of similar on similar, in a way that is most often beneficial for them. All that is red helps the blood not to abandon the body, it assists the blood, saves the body. I have proposed assigning the adjective active to this second type of analogy of expansive thought. Of course, I shall not carry on with the innumerable complications introduced in this thought by the reactions between these analogies, or by the items of our intuitive and common knowledge of the world that they express or imitate by giving birth to a sort of universal vitalism; I shall stop here because I have too much to say.

Just one more observation, however: if expansive thought began with human intelligence, if it underpinned our first generalisations, our first beliefs, our first activities, we have to find it in its pure form in the rare people who have not been touched by civilisation's progress. And it is that which the great work of Mr Lévy-Bruhl[3] on *primitive mentality* defines, describes, and at times resurrects. It is in some ways active analogy that allowed for the admirable schematism that the author called *law of participation*, which he defined as follows:

> I should be inclined to say that in the collective representations of primitive mentality, objects, beings, phenomena can be, though in a way incomprehensible to us, both themselves and something other than themselves. In a fashion which is no less incomprehensible, they give forth and they receive mystic powers, virtues, qualities, influences,

which make themselves felt outside, without ceasing to remain where they are.

And yet we saw earlier that, for a sixteenth-century scholar, everything that was red would heal a bleed and thus save the body. Mr Lévy-Bruhl, whom I can only quote, discovered that, for a similar reason, the indigenous peoples of Australia often covered their bodies in a layer of red ochre; and this was not purely ornamental. Red ochre gave them a beneficial and effective strength: 'it is an equivalent of blood, it is blood' ... Of course, I am not about to compare the psychology of a naïve savage with that of a sixteenth-century scholar, who was energetically releasing to his delight a form of spontaneous mind that had been annihilated, oppressed, and repressed by scholasticism in decline! But, finally, was the development of expansive thought not completely clarified by Mr Lévy-Bruhl's work? We thus understand that for this type of thought, a concept, or rather a word, cannot be fixed *a priori* to a strict definition, because this concept or word is a central source of impetus for the intelligence.

We shall not go as far as Mr Lévy-Bruhl by judging as 'pre-logical' and 'mystical' expansive thought, which lacks the scruples of formal logic, which overcomes contradictions by its rapid pace, which worries little about distinguishing between dream and reality, which impedes the mind even more than it enriches it with a mass of diverse, unanalysed, and unverified notions, and which never, furthermore, uses the slightest precautions against the equivocations that it engenders. Rather, we shall say that expansive thought is entirely unreflective. But how does one go from unreflective thought to the reflective thought of which we shall briefly study the *a prioris*? Descartes answered this question at the beginning of his *Principles* with a few lines of introduction that I am sure you all remember, but to which I ask you to indulge me in listening:

> As we were at one time children, and as we formed various judgments regarding the objects presented to our senses, when as yet we had not the entire use of our reason, numerous prejudices stand in the way of our arriving at the knowledge of truth; and of these it seems impossible for us to rid ourselves, unless we undertake, once in our lifetime, to doubt of all those things in which we may discover even the smallest suspicion of uncertainty.

We then move from *expansive* thought to *reflective* thought just as we move, as far the mind is concerned, from infancy to adulthood.[4] Reflective thought takes off when it experiences the need to possess knowledge and to be in control of itself, when it manages to distinguish truth from falsehood, when it verifies the materials to be used to construct the theoretical monument of science. Reflective thought is first characterised by its constantly polemical attitude. By placing its trust only in its own judgement, reflective thought says 'no' to all that appears obscure, unfounded, and unbelievable. Its positive conclusions are always the result of a refusal; in their original and effective form, its *a prioris* always follow from a defence against error and superstition.

What is immediately clear to those who study reflective thought is that it is in some way opposed to expansive thought, it disavows the naïve beliefs of expansive thought, it mocks expansive thought's extraordinary assertions. But they also see that reflective thought contains the impetus of expansive thought, even if it alters its direction, even if it moves towards a diametrically opposed spiritual horizon. This is what should be established; but I must restrict myself here to a few brief remarks.

A moment ago, I said that the *a prioris* of reflective thought are always the result or the other side of a negation, as it were. I should therefore demonstrate to you what, from this point of view, I mean by what has been called the demands of rational thought. I shall explain this by using just two examples, reminding you that today is all about the science of material objects.

Now, material objects apparently have quite diverse properties: they are hot or cold, dry or wet, solid or fluid, brightly or darkly coloured etc. In addition, they do not remain similar to themselves, they are changed by fire or mechanical actions, and they react with one another in often quite unpredictable ways. The mind, astounded by this great richness and, as if driven mad by the exuberance of nature, which has presented it with so many unbelievable stories, at some point starts to distrust the senses and declares, once and for all, that the apparent and ostensible qualities only result from the 'opinion' of our body. Outside us, there is only one material substance that is absolutely without quality and that has no other property save from occupying a certain place. The scholar will admit that bodies are formed from infinitely small fragments that all resemble this substance, fragments that, by their movements and reciprocal reactions, produce the world such as it appears to us. And from this denial of qualities immediately stems the strictest form of atomism, whose influence over the forming of scientific theories is known... Yet, the atomist hypothesis, which according to Hannequin 'is a necessary hypothesis that stems from our very knowledge', does not satisfy everyone.[5] Even Hannequin admits that it contains irresolvable contradictions, for reflective thought asks straightaway: if in a manner of speaking atoms float in the void, why is there a difference between the space occupied by the atom and the place where the atom is not? Furthermore, intransigent rationalists such as Descartes strove to explain the universe by 'bare extension and its modifications'. If the early nineteenth-century chemists sometimes loathed to follow the atomist hypothesis, it was because it contained, according to them, discontinuities that experience did not demonstrate, and that reason saw as having no purpose. However, this discontinuity, produced by reflective thought, which, by accepting it, perhaps relied on the tendencies of expansive thought, would itself produce the idea that 'thou [God] hast ordered all things in measure and number and weight'. And it was in an inspiration similar to that which characterises this phrase from the Wisdom of Solomon that Berzelius liked to quote, that Berzelius himself saw [as] the cause not only of Dalton's renewal of the atomist theory, but of the whole discovery of the law of definite proportions.

Here is a clearer and less abstract example: was the experimental method itself, with its measuring instruments and elaborate procedures of verification, not borne out of a lack of trust in everyday experience, in what was well known, in the 'everyone knows', and in the authority of tradition? This is at least what seems to stand out from Bernard Palissy's admirable dialogues,[6] which I recommend you to read again. There we listen to the credulous, expansive, enthusiastic Theory narrating incredible, extraordinary, unbelievable experiences, allegedly made in the laboratory. In contrast sceptical, reflective, critical Practice – which represents the rise of the modern mind – by relying on *a prioris* which stem from a sound judgement of negation, refuses to believe the experiences of Theory, that is misleading or was misled.

Let us bring to an end this far too brief attempt to demonstrate by means of example. I should like to talk to you on another occasion about the reasons for our reluctance to accept an action between distant objects without an intermediary which propagates this action. I should like to talk to you on another occasion about the tendency towards identification, which the much-missed Meyerson[7] saw as the only *a priori* brought about by our intelligence, indeed the only ostensible characteristic of reason itself. Finally, I should talk to you about the mutual responses of expansive thought and reflective thought in the development of current scientific theories ... I have no need to apologise for not being able to say everything. I will only note that, from the point of view that we have adopted here today, it is totally wrong to insist with Mr Rougier[8] that rationalism contains a hidden mysticism; it is totally wrong to speak, along with Mr Rougier, of the 'paralogisms of rationalism'. Allow me to explain by means of an assertion from outside of science, which will shorten the demonstration. Mr Rougier says that the declaration 'all men are equal' is an unjustifiable dogma that is not based on any judgement. We can now respond that such an assertion is the simple response to a negative judgement. Seeing social inequality, the philosopher asks: 'Why are there so many differences between men; why are some blessed and others miserable?' And, after consulting his conscience, the philosopher replies that these inequalities are not forcibly imposed on reason, that they are unjust, irrational, to be corrected. Or do you instead see in there paralogism, or irrational and pure mysticism?

Let us finally tackle the question you are dying to ask: 'Can human thought occur without any *a priori*?' You know that the positivists hoped as much for the sake of objectivity, you know that the supporters of scientific nominalism thought they had reached this conclusion, you also know that today the members of the unruly Vienna Circle want it to be so with exuberance and juvenile aggression.[9]

The scholars who belong to this Circle profess the most absolute experimental empiricism. They explain this empiricism using verbal resources offered by logistics and new logic; they declare that they will succeed in solving all the problems to be solved by relying on protocol statements or reports of their laboratory work. Furthermore, they say with great confidence that no one has the right to ask them the questions that they do not ask themselves, that these questions are absurd, devoid of meaning and without any possible intelligibility; finally, they claim the disintegration of the *a priori*.

Here is an adversary that has been beaten, bound, and imprisoned before it has even fought. What is it to do? Since its restraints prevent it from taking on the theory in hand-to-hand combat, it can only try to see whether the Vienna Circle has not first killed within itself the expansive thought that remains a source of inspiration, even when it is educated and disciplined by reflective thought. It can then try to see if the Vienna Circle has not killed within itself the reflective thought, which, on every occasion of doubt, goes back to its nascent state to pronounce a judgement. Finally, it can investigate whether the Vienna Circle, which thrives on a verbalism no doubt rigorous, but not invigorated by thought, does not risk returning to the most barbaric of scholastics.

From the prison in which it has been shackled, it will thus proclaim with a strong voice that human intelligence produces light at the same time as being an organ of vision, just like the phosphorescent eyes of deep-sea fish; if one prevents it from providing its own brightness, it quickly becomes blind. The historian of science will conclude that no method, no verbal or industrial method, however sophisticated, can be an effective substitute for thought and the continuous awakening of the mind.

## Notes

1   Talk delivered at the Unit for the history of science of the Centre international de synthèse on 20 November 1935.
2   I have studied virtual analogy and active analogy in a small book on *Les concepts scientifiques*, Alcan 1926. Mr Cresson has shown the role of analogy in his small book *Les réactions intellectuelles élémentaires*, Alcan.
3   See *Archeion* 12 (1930), p. 15: 'La philosophie de Lucien Lévy-Bruhl et l'histoire des sciences'.
4   In his excellent book *Les âges de l'intelligence* (Alcan 1934), Mr Léon Brunschvicg makes a similar remark from a different point of view.
5   Essai critique sur l'hypothèse des atomes.
6   Mme Bessmertny will give a talk on Palissy's dialogues.
7   See *Archeion* 11 (1929), p. xxxiii: 'La philosophie d'Émile Meyerson et l'histoire des sciences'.
8   *Les paralogismes du rationalisme.*
9   The publisher Hermann has published a series of works in French translation of the main members of the Vienna school in order to advance their knowledge in our country. See *Archeion* 17 (1935), pp. 332–333 and 18 for reviews; see also *Archeion* 17 (1935), p. 421: 'Réflexions sur le congrès de philosophie scientifique' (Paris, Sorbonne September 1935).

# Appendix 4: The philosophical method in the history of science[1]

I – Before even hearing my talk on the subject that, following [Henri] Berr's wish, I would like to submit to your reflection, I fear that some historians of science will raise the following preliminary question about the topic of my talk, regardless of its content: 'is there such a thing as a philosophical method in the history of science?' And, without further ado, without further consideration and as a matter of course, they will answer their own question thus: 'There is no philosophical method in the history of science. It is not possible to have a philosophical method in the history of science. Science and philosophy should not interfere with each other; they are not concerned with the same things. The historian of science does not have a valid excuse if he lets himself be seduced by anything outside the scientific domain; he must know science and he must know general history in order to be able to accomplish his task. Equipped with this indispensable prior knowledge, he must refer only to texts. He must positively demonstrate that this scientific advancement was made in this year by the effort of this scientist; that this idea came for the first time to this researcher under these conditions; that this experiment was performed for the first time in this place and in this way ... Any attempt made by the historian to move away from the solid ground of facts is clearly reprehensible and chronological empiricism, which is all that our efforts can achieve, should be the ultimate goal of his effort. If, however, a historian wants to penetrate completely the attitude of the past scholar, whose work he is studying, if he wants to try to ascertain what the deep motivations that brought the theory to life were by hypothetically penetrating beneath the surface of these theories, then we warn him that he is going off-piste and we call him to order'.

I shall not respond for the time being to such criticisms, which, if accepted as valid, would irredeemably condemn all the work that I have undertaken, and that I was under the illusion of doing quite well. To my opponents, I shall only say that if it had been demonstrated that the history of science is only able to satisfy a curiosity that is of course legitimate but philosophically sterile, [a curiosity] that delights in scholarship, picturesque events, elegant patterns of descriptions of theories, [and] from which the creative thought has effectively disappeared ... well then I would immediately cease to devote myself to the history of science. So, you are going to say to me: what do you ask of the history of science? I hope that, little by little, I shall be able to make you understand it.

First, I should make clear that when I speak of the history of science, I am talking about the history of scientific thought and nothing else. The other aspects of science, including observation, experimentation, measuring, calculation procedures or techniques for building laboratory equipment, either have no importance, or are involved only as auxiliaries, as an aid to thought or the creation of thought. You may be surprised at such a manner of speaking, and you may find it dreadfully perfunctory, to put it mildly. I shall try straightaway to ease the bad impression you may have by confessing that, in my opinion and to cite a specific example, all the reagents that one encounters in the labelled bottles on the shelves of chemistry laboratories, all the instruments that one finds in these same laboratories are embodied products of a theory, which are used in order to be able to verify this theory, but which must be understood and in any case can only be understood in relation to this theory. In other words, laboratories do not date back to the creation of the world; they do not pre-date the development of all chemical theory as the work of generations of investigators guided it. There is a fine book waiting to be written both on this topic and on the parallel progress of experimental technique and theory, the title of which, to borrow from Mr Brunschvicg, could be *The progress of experimental consciousness*.[2] I do not have the audacity to venture into such a challenging work, which I hope will come to fruition; but this is not our topic today, so let us move on.

II – The history of scientific thought cannot do without a sufficiently large and, if you permit me to say, sufficiently cumbersome learned apparatus. [However], forgive me for insisting, scholarship is not the aim, but rather the means and tool of the history of scientific thought. As an indispensable tool, it must be honestly acknowledged that scholarship must be constructed as carefully as if we had the hope that it would provide us with real knowledge all by itself, just as the pseudo-scholars that Malebranche mocks so subtly did. But this hope fails us entirely. Thanks to the aid that scholarship gives us, thanks to the thoughts it inspires within us or that we apply to it, we shall aim: 1) to understand the human mind better; and 2) by means of this very knowledge, to use our intelligence more wisely and less empirically than we have done thus far, building as we have done our scientific, philosophical and historical theories haphazardly. Or rather, without deceiving ourselves with grandiose projects or long-unfulfilled dreams, we would like to manage to provide some services to scientists and philosophers.

III – 'Your attitude', you will say, 'is entirely *a priori* and in fact precedes all historical work'. I agree. I would respond that this attitude dictates, broadly speaking, the working method that I employ. I kindly warn you that I cannot prove to have undertaken my work in this way, and that you will be able to pit one theory of the history of science against another, one method against another and one set of results against another.

IV – I shall now ask you to accept, as a premise that is impossible to verify and impossible to avoid, that even if the human mind is always and everywhere similar in its fundamental characteristics, even if it truly has an unalterable framework, the attitudes that it is able to adopt and that effectively determine the orientation of men's mentality are diverse and highly variable. [I shall also ask you to accept] that one must see the principal source of the heterogeneity of opinions professed by the diverse investigators precisely in this heterogeneity of orientations of mentality.

V – I dare to hope that the history of science will enable us to construct a catalogue (it goes without saying that it would be incomplete) of these different possible attitudes, and that, with the help of this catalogue, it will enable us to create a useful classification of a certain number of hypotheses that appear in almost identical form throughout the sciences and the various eras of human development.

VI – But having an abstract and schematic insight into the various frameworks into which humans have attempted to draw in the various aspects of the world is not enough. One must go further than listing all the possible hypotheses and discerned classifications. The historian of science must have, early on in his work, the will to use the writings that he studies to endeavour to capture thought in its nascent phase, thought in the moment that it is formed within us, thought that arises in the thinker at the precise moment when, in a manner of speaking, it awakes within him. I use the expression 'thought in its nascent phase' in the same way that one talks of the nascent state of hydrogen in the laboratory, active hydrogen that frees itself instantly when hydrochloric acid attacks zinc, as opposed to the less active hydrogen stored in an ordinary container. This is only an image, which I hope will be clear to you.

I grant you that it is not easy to reconstruct thought in its nascent phase – active thought, thought that is at the same time created and creator, that was the origin of all progress in scientific theory – in our own minds (and the only reagent available to us is our own mind). I grant you that such an effort comes with risks and that our conclusions will be subject to discussion. We are no longer on solid ground and we are at risk of getting bogged down in interpretations and opinions; some philosophers and some historians of science could, with reason, dispute the solidity of our interpretations or contest our opinions. No matter: if we are confident about the validity of our scholarship, are we going to refuse to take a risk that appears unavoidable? Here I must speak for myself, as perhaps you have faculties that I lack: I confess that I cannot understand a text without grasping it from within, without interpreting it a little. I know full well that the method I am recommending has been essentially condemned by a philosopher for whom I have always professed great amicable admiration, the late Émile Meyerson. I know full well that, in order to be certain of avoiding a surreptitious introduction of one's own mentality into that of the past thinker, and in order to avoid all 'introspection' (this is the term he himself uses),

Émile Meyerson asked, in a fine 'study on the products of thought',[3] that one study these products as something entirely external to the judgement of the historian-philosopher, as a purely inert object. Such an attitude would be perfectly positive and could enable us to construct a blueprint of the evolution of theories, even to perform a critical analysis of these theories by systematically ignoring thought in its nascent phase, which we can only grasp when it is produced within us. This attitude, which is not our own, is legitimate if its aim is to explain science and its history in accordance with the data of common sense that are assumed to be fixed and intangible. It leads to an epistemology that perhaps does not have the universal impact that its author attributed to it, but which defines a view that is widely held. One might admit that thought, to the extent that it is separated from its nascent phase, is transformed into a thing and is inserted into common sense. We should note, however, that if we wanted to follow Meyerson's advice all the way we would not stop as he does at the stage of common sense [as] we would soon not understand anything anymore. After all, if the words employed by a particular past scholar mean something to us, this is due to the fact that we pour our thought into his and that we have decided to speak his language when we think we understand it. Here is an insurmountable barrier for any working method that, due to a lack of trust in ourselves, imposes a strict nominalism; any scientific effort to bear fruit must rely on a minimum level of reality penetrating our minds. Boerhaave had already considered this limit when he intended to study fire without even knowing what fire was and refusing to accept any prior notions. If our ignorance were complete, we would reject even the discernible experience that served as our starting point, and the word 'fire' would have no more meaning for me, a physicist, than it would for an Indian or an African. We cannot reject our experience and thought entirely. But if we do, we must resign ourselves to keeping quiet. In order to construct a theory, one must have a point of departure. And this arbitrary point of departure casts a worrying shadow of doubt over the entire theory, which [for this reason] will be subject to new examination and revision. Why [then] would the historian of scientific thought refuse to run the risk that the scientist must accept and keep in mind each time we witness a revolution in theory?

VII – Thought in its nascent phase is therefore something entirely other than the thought that has entered history. The thought that entered history undergoes a devaluation, a levelling out, an erosion, the final stages of which are Poincaré's conventionalism and Meyerson's identification. This must be established by theory and verified in the facts by the history of science. I urge you not to be too critical of this statement, seemingly prior to history, which appeared to me in all its generality upon contact with the texts that I studied. We are here in the midst of the philosophy of history. Although I abhor authorities and although I am only speaking for myself, allow me to have the vanity, or weakness if you prefer, to invoke for a moment the great Montesquieu, who defined his method in the preface to

*The Spirit of the Laws*: 'I have laid down the first principles, and have found that the particular cases follow naturally from them'.

I therefore assume *a priori* what I accept to be true, and verify it if I can; I make no distinction between a working hypothesis and a realist supposition which, until the evidence shows that it is false, amounts to a conviction. I recognise without any humility, and because I am a human being, that I may make mistakes, and indeed that I have made many mistakes; I then correct my error without sustaining a blow to my pride. Those historians who reject my approach are free to criticise me and improve upon me by proceeding differently.

VIII – Here is then a problem that I must introduce because it affects the theoretical foundations of my method. However, I can only but state it because it transcends the history of science. How can a theoretical monument created by the activity of thought in its nascent phase be turned into an instrument, even a passive instrument, that one can either use or replace with another?

IX – Here is a similarly significant problem that I only mention in passing: how can active thought, thought in its nascent phase, thought that created a theory, let itself be enclosed passively into an account of phenomena?

X – Despite the creative activity of thought in its nascent phase, the historian can discern its origin, milieu and time; he can date it. He can say: this thought was born in this country, at this time, and it bears the hallmark of the social facts and the philosophical and scientific styles of the time when it appeared.

XI – What is the role of the milieu and the time, to which Taine drew attention and which he believed to be entirely decisive? We shall say first of all that they are simply inhibiting. Only those theories driven by trends that their milieu and time do not paralyse or even do not annihilate manage to become established or achieve success.

XII – Milieu and time also play a significant role in directing, strengthening and amplifying theories that are driven by the same trends as their environment. They act through education, the press, customs and traditions, communicating this or that orientation to minds. There is a wide variety of actions and reactions here that we highlight without attempting to analyse them, and, as politicians do, we call them imponderable.

XIII – It must be clearly pointed out, however, that thought in its nascent phase, despite being under the influence of its milieu, and often receiving its attitude and orientation from the trends of its time, is not a direct product of its milieu and time. Contrary to what one might believe, when one studies its schematic evolution, it is not *de jure* a prisoner of its milieu and time, although one could argue that *de facto*, it depends a lot on the circumstances of its first appearance. Furthermore, as the properties of the milieu and time have retained a certain plasticity and are themselves changeable, thought in its nascent phase can be counted among the factors acting on them by transforming them little by little.

Hence this double obligation for the historian of science: 1) to make one-self the contemporary of the scholar whose work he is studying, and employ to this end all the means that the most careful scholarship affords his active sympathy; and 2) to resurrect and update the theory that he is presenting to the public as if it had no context, and as if it were I dare say independent of time and space. How can this second task be resolved at the same time as the first? Let us leave this issue of method, which in fact exceeds method and on which our reflections will shine a glimmer of light.

XIV – The scientist, by looking to express symbolically or theoretically the truth about the facts that he studies experimentally, introduces an element of finality into the scientific theory itself. As such, this element is entirely foreign to the phenomena that he is seeking to link in harmony using constant laws. Similarly, by looking to capture thought in its nascent phase, the historian of science introduces into his work an element of finality which, as such, appears foreign to the perfect understanding of these texts.

We need to distinguish the theory such as it emerges in the mind of the person who thinks it from the thought that becomes a thing, an object, and that becomes stabilised; to achieve this goal – it is necessary to repeat this constantly – we have to recreate the theory, with all its creative effectiveness, within ourselves.

XV – The result of all of this is that the intelligence of the historian must not begin and end with the 'objective' reading and interpretation of the text, which gives the outward appearance of thought in its nascent phase, rather than the outpouring of this thought itself.

In other words, from the point of view of the philosophical method in the history of science, the reading and understanding of texts dated and situated in their historical milieu is only a brief stage in the work of the historian; the reading and interpretation are necessary but insufficient.

XVI – The history of scientific thought is a creation of the mind, definitely not *ex nihilo*, since it is constructed with texts that are solid matter, but a creation nonetheless.

I shall not dwell on this, and I shall simply tell you that the investigation of documents and the understanding of texts that are done previously to or at the same time as the construction already presuppose a certain choice or, if you prefer, a certain arbitrariness. However, it does not follow that the history that we create or re-create is just a story, and that the inevitable partial arbitrariness is just a whim or fantasy on the part of the historian. You have understood that, on the one hand, any construction can be altered, demolished or remade when our investigations or reflections progress, and that, on the other hand, there is a certain hesitation, which is inevitable when interpreting any writing, speech or any human action.

XVII – Just as the scientist can give various interpretations of the facts that he has decided to study, likewise the historian of science can give diverse interpretations of the texts that he has decided to use in his construction.

If a text supports several interpretations, it is not impossible that these interpretations had actually been given by different commentators on the same scientist. Indeed, it is often the case that the evolution of theories moves forward due to this dissociation of fundamental doctrines.

XVIII – The historian of the sciences, while clearing the path before him by a *quick* critical analysis of the texts that he already knows and that have provided the foundation of his reflection, must also quickly read the greatest numbers of similar texts of the same period:

1) to ensure that he has not made a serious mistake in his interpretation and if need be, rectify it; 2) to absorb better the fundamental attitudes that determined the tendencies of the mentality of the authors under study; 3) to be able to provide a general framework containing each of these attitudes and tendencies of mentality as a particular case.

XIX – In order to carry out such a piece of work successfully, the historian of scientific thought must know first of all that the problem of the genesis of hypotheses is entirely independent from the problem of the verification of these same hypotheses. This verification, it should be noted, happens at a different stage of thought and work. The confusion between these two heterogeneous problems could prevent one from grasping the theoretical content of scientific writings, inextricably mixing the theory with the facts.

Let us cite just one example. The supporters of the *natural philosophy of metals* claimed that imperfect metals such as lead, iron, tin, copper, quicksilver and silver itself have a natural tendency to transmute into the perfect and incorruptible metal, gold. One would not understand how they could have been led to produce such a hypothesis if one looked for the origin of this oh-so-seductive hypothesis in the experimental facts concerning the metallic realm.

XX – A notion or hypothesis can be transported directly or analogically from one class of things to another, and from one science to another.

There are so many examples of this that a large tome would not be enough to list them all. Thus, the *natural philosophy of metals* of which I was just speaking was certainly inspired by a biological analogy. One finds constantly in the writings of its adherents that 'the imperfect metal is to the perfect metal what the child is to the adult, what the green fruit is to the ripe fruit'. A child grows, a fruit ripens; this is what suggests the idea of perfecting, the idea of progress, the idea of evolution. One could also add that this is what suggests the law that I have learnt in school, and that, considering the human life span, can only be a boldly extrapolated analogy: 'every living being goes through the same phases of development as those through which the evolution of its species has travelled throughout the ages'. Another thing: if there is a class of bodies that appears to have been established without debate by our spontaneous experience, it is the class of metals of which I have just spoken. They have a higher density, a remarkable shine, and they have a family resemblance. How does one justify that these bodies form a class, if one rejects (and this is possible) the biological analogy that I have just mentioned? We could, for instance, form a substantialist hypothesis and say that metals are composed of two bodies: one which characterises the type and which is common to all metals, i.e. the principle that makes them metals, the other which is particular to each metal and makes it possible to name it gold, silver, copper etc. Well, this opinion, which was Boerhaave's, could not persist in science; the metals systematically refused to let themselves be split into their hypothetical constituent parts and they persevered in remaining simple bodies. But we have transferred the substantialist decomposition, resisted by the metallic kind, to other kinds. Lavoisier was able to assert that acids are composed of two bodies: the first is the acidifying principle, which is oxygen and which made it possible to name them acids, whereas the second is particular to each acid, making it possible to name each individual acid, for example sulphuric or carbonic acid.

XXI – The relationships between the classes of bodies isolated artificially and the hypotheses justifying the creation of these classes are not fixed. They are partly arbitrary and are likely to vary; the repercussions of this arbitrariness extend to the evolution of the theory, which can successively try out several heterogeneous and incompatible relationships, splitting up some classes and creating new ones. From this we can conclude 1) *n* heterogeneous hypotheses can correspond to a single notion (or to a single class of things); and 2) *n* notions or classes of things can correspond to a single hypothesis or one type of hypotheses.

XXII – A single hypothesis can be supported by a number of incompatible explanatory justifications, and the scientist who critically analyses his own thought is obliged to choose between them. We should note, however, that this choice, which is imposed by an aesthetic desire for harmony within the theory, is not the result of a spontaneous movement of expansive thought; it is rather the result of a desire for clarity and is dictated by reflective thought's principle of non-contradiction.

This is very significant for the historian of science; because if we accept, to give just one example, that imperfect metals have a natural tendency to evolve towards perfection, a number of procedures can be used to explain this hypothesis. [Here are a few]: a metal is always the representative of the same being that we know at various stages of its development; or the perfect metal, gold, is fed by imperfect metals which, by means of assimilation, are transformed into its own substance; or even the transformation of imperfect metals into gold continues over time. [Other explanations could be the following]: this transformation would occur instantly if nature did not encounter some resistance that would prevent it from acting according to its own tendency; or it is possible to transform imperfect metals into gold, whereas the contrary is impossible and against the order of things. To these explanations for the hypothesis, one might add other explanations that do not contradict it but that do not support it either: one could say that there is but one matter and consequently that anything can be transformed into anything. One [could offer] a substantialist hypothesis etc.

What is remarkable is that these incompatible explanations are involved in the very same texts and if we do not conduct the analysis (an extract of which I have just outlined to you) before reading them, these writings would be totally incomprehensible. We can thus understand the role of what Mr Lévy-Bruhl has called *prelogic* by studying the primitive mentality, which in short is nothing but expansive thought forging ahead. We can thus understand the attitude of scientists who do not hesitate to employ incompatible hypotheses as tools, after denying that these hypotheses are of any value for the penetration of reality. Let us recall again that Duhem, who also refused to accept that a hypothesis represents reality, protested against the various interpretations of a single hypothesis, admitting at the same time that he had no logical reason to oppose those who accepted the scandal (his use of the word) of logical contradiction.

XXIII – The explanatory justifications of very general hypotheses can themselves be considered to be hypotheses, if the main hypothesis is elevated to the rank of fact; scientists spontaneously move from one way of seeing to the other.

Let us clarify these transformations in attitude by means of an example. For 18th century physicists, the hypothesis that light is a body remains one of the suppositions that were currently unverifiable. By contrast, for the chemist, this hypothesis, which explains so many phenomena, became an evident fact. First of all, and in the context of the Newtonian system, it was proclaimed that the matter of light is like any matter that is subjected to the forces of inertia and universal attraction. It was then claimed that, as with any chemical reagent, light has specific attractions, which are already manifest when it enters one body from another by refraction. It was asserted that, in the most common cases, the refractive index of a substance, regardless of what the substance is, is directly proportional to its density. It is proclaimed that if a body is combustible, that is to say its constituent part is the matter of light, its refractive index is greater because light has more affinity for a body that contains light than for bodies that do not contain light.

Diamonds fall into this category, yet no one had ever burned them. When the experiment was finally attempted, when the experiment succeeded, scholars such as Buffon claimed that the mental eye had pre-empted the laboratory procedure, and that this costly procedure was practically pointless. Who would have doubted the corporeality of light faced with such success? However, at the end of the 18th century, Lavoisier's chemistry, without in the least attacking the theory of the corporeality of light, did not know what to do with light and was content with poetically thanking the benevolent

God who permitted us to see the day. This silence of chemical theory did not prevent Biot and Arago from writing the following in 1802:

'The physicist who observes the refractive powers of substances to compare them, acts like the chemist who presents successively the same base to all acids, or the same acid to all alkalis ... In our experiments, the substance that we presented to all bodies was light'. Biot had hoped that the refractivity of bodies would reveal the composition of different reagents to the scientist without him having to perform a chemical analysis. He failed; early 19th century chemists, resigned to not being able to find even the slightest link between bodies' refractivity and their composition, ended up abandoning [the hypothesis of] the materiality of light, and coming around to Fresnel's theory, which drove it out of chemical doctrine. Here is a chapter in the history of optics that I am studying with Madame Bessmertny, a chapter which is (unless I am mistaken) unknown, and to which philosophical reflection has led us. After all, one simply needs to search for the numerous texts on this problem in order to find them; and one would only search for them because of a prior reflection of the sort that I have just discussed.

XXIV – The fundamental notions upon which science relies at all times for its development are not intangible data. Although some of them were dictated in some way by our knowledge of the empirical world, they are, to a certain extent, plastic. They can be modified in order to be harmonised with our entire knowledge. Conversely, they can modify our spontaneous world system.

XXV – I think I have given you a general overview of the philosophical reflections that have asserted themselves imperiously over the course of my work, and that have guided my work. I have left out the effort of the construction and composition of the history of scientific thought, as it would be beside the point. You now know that I have always tried to put the potential reader in contact with thought in its nascent phase (I am speaking for myself so as not to commit others), either in its expansive form, which forges ahead spontaneously in all directions where it can find a way through, or in its reflective form, which harshly criticises itself and denies all that appears to be incomprehensible or illogical. You know that I would like this thought in its nascent phase to be reconstituted in the mind of the reader, who must for a brief time be the disciple and contemporary of the past scientist, who confides in him through my pen. You know that from this point of view, the researchers' efforts, whether ending up in failure or resounding success, must be placed on the same level, as failure and success do not directly depend on human intelligence, but on the contact between man and the world.

I shall add two further points. The first is that the philosophical method must remain completely behind the scenes of the history of scientific thought, and that the reader (the only final judge of any book) must not be forced to think about this method. A work that draws attention to the author's efforts instead of just casting light on the activity of past scientists is not quite right. The second point is connected with all the reflections, the conclusions of which I have outlined to you too schematically. There are perhaps philosophers and historians who will reject them entirely; I have nothing to say to them; let them fight me. There are perhaps others who will accept them either partially or entirely but who will criticise me anyway by asserting that they had come to these conclusions or that they could have come to them directly by means of a well-constructed dialectic that did not require a detour via the history of science. They could therefore ask how the history of science can help us to understand the human mind better. I do not need to respond to these people. It is the work of historians of science that must advocate the history of scientific thought. I shall therefore be content with saying again that if the historian of science is able to bring some valuable tools to epistemology, psychology, philosophy and science, tools that they could make use of, this would be the highest accolade available to him.

## Notes

1 Talk delivered at the Unit for the history of science of the Centre international de synthèse, on 27 April 1937.
2 [This imaginary title is modelled on Léon Brunschvicg's *Le progrès de la conscience dans la philosophie occidentale* (Paris, Alcan, 1927)].
3 [Metzger refers to Émile Meyerson, 'De l'analyse des produits de la pensée', *Revue philosophique* 118 (1934) 135–170, although she does not cite the precise title; she rather mentions his 'études sur les produits de la pensée'].

# Index